This item must be returned or renewed by the last date shown below. The loan period may be shortened if it is reserved by another reader. A fine will be due if it is not returned on time.

DATE OF RETURN

STEM CELL RESEARCH PROGRESS

STEM CELL RESEARCH PROGRESS

PRASAD S. KOKA

EDITOR

Nova Science Publishers, Inc.
New York

NOTICE TO THE READER

LIBRARY OF CONGRESS CATALOGING-IN-PUBLICATION DATA

Available Upon Request

ISBN: 978-1-60456-308-5

Published by Nova Science Publishers, Inc. ✦ New York

CONTENTS

PREFACE

The two broad categories of mammalian stem cells exist: embryonic stem cells, derived from blastocysts, and adult stem cells, which are found in adult tissues. In a developing embryo, stem cells are able to differentiate into all of the specialised embryonic tissues. In adult organisms, stem cells and progenitor cells act as a repair system for the body, replenishing specialised cells. As stem cells can be readily grown and transformed into specialised tissues such as muscles or nerves through cell culture, their use in medical therapies has been proposed. In particular, embryonic cell lines, autologous embryonic stem cells generated therapeutic cloning, and highly plastic adult stem cells from the umbilical cord blood or bone marrow are touted as promising candidates. Among the many applications of stem cell research are nervous system diseases, diabetes, heart disease, autoimmune diseases as well as Parkinson's disease, end-stage kidney disease, liver failure, cancer, spinal cord injury, multiple sclerosis, and Alzheimer's disease. Stem cells are self-renewing, unspecialized cells that can give rise to multiple types all of specialized cells of the body. Stem cell research also involves complex ethical and legal considerations since they involve adult, fetal tissue and embryonic sources. This new book presents the latest research from around the globe in this dynamic field.

Chapter 1 - Hematopoietic stem cells (HSC) are associated *ex vivo* with a cell subset actively rejecting lipocationic dye such as rhodamine123 (Rho) in the extracellular medium via the multi-drug resistance gene product. It is however not known whether this association remains after culture. The authors therefore evaluated whether cultures derived from CD34+ human cord blood (CB) cells contained dye-excluding (Rho-low) cells enriched in early HSC. Our investigations showed that after short-term culture with FLT3-ligand, thrombopoietin and stem cell factor, Rho-low cell were still detectable, but with a lower frequency than that of uncultured cells. Rho-low cell frequency varied depending on the stimulus used *in vitro*, and these variations correlated with that of long term-culture initiating cell (LTC-IC) frequencies ($R=0.619$, $p<0.04$), whereas no correlation could be established with late HSC (CFU) frequency. Rho-low cells sorted after 6 days of culture contained 2.8-fold more secondary CFU and 1.8 more LTC-IC compared to Rho-high cells. Though these data suggest that dye exclusion assays could be used as a relevant alternative compared to other *in vitro* assays to identify immature HSC after culture, it remains to be determined whether Rho-low cells derived from cultures retain clinically relevant characteristics such as the ability to reconstitute a lethally conditioned host. In this light, the authors further discuss below the limitations of some *in vitro* and *in vivo* hematopoietic assays, and their relevance as tools to identify HSC that are able rescue conditioned patients.

Chapter 2 - *Background:* Clinical trials are ongoing involving transplantation of ex vivo expanded hematopoietic stem cells. This study was carried out in order to better understand the effects of ex vivo expansion on the rheology of umbilical cord blood-derived stem cells.

Methods: Micropipette aspiration and recovery experiments were performed using fresh umbilical cord blood CD34+ cells and umbilical cord blood CD34+ cells that were cultured for 1 week. From these experiments, viscosity and cortical tension were determined, and results were analyzed to determine if cultured cells differed from fresh cells in terms of rheology.

Results: As compared to fresh cells, cultured cells had a significantly lower viscosity (221.41±16.88 Pa•s as compared to 515.42±62.87 Pa•s for fresh cells, p=0.0001) and a significantly lower cortical tension (5.54±0.49x10-5 N/m as compared to 2.14±0.23x10-4 N/m for fresh cells, p<0.0001).

Conclusions: Cultured and fresh CD34+ cells are likely to behave differently in the body. Cultured cells are likely to flow more easily through the capillaries; however, they may also recover more slowly from a deformation.

Chapter 3 - Recent studies indicate that a variety of cells derived from adult human tissues show the capacity to change their tissue specific differentiation program. In this study the authors harvested cells from lipoaspirated adipose tissue, trabecular bone and pericranium tissue in order to evaluate and compare the differentiation potential of these populations by exposing them to adipogenic, chondrogenic, osteogenic and neurogenic induction media. Histological and immunochemistry analysis after adipogenic induction revealed adipocyte-like characteristic for all populations. Under osteogenic induction, all populations expressed the bone matrix proteins osteocalcin, osteopontin, osteonectin and secreted alkaline phosphatase, an early bone marker, into the culture medium. The chondrogenic potential was evaluated after staining of acidic glycoconjugates within the extracellular matrix. All induced cell populations were stained positive, indicating the presence of *de novo* cartilage. After neuronal induction, responsive cells exhibited neuron-like morphology with refractile cell bodies, extended long processes terminating in typical growth cones and long axon-like filopodia. Immunocytochemistry revealed the expression of neurofilament M and neuron specific enolase. Due to the use of heterogeneous cell populations each cell population preferentially expressed the phenotype of the tissue it was isolated from. In comparison; under the respective experimental conditions the periosteal and adipose derived cell populations demonstrated equal versatile developmental potentials according to cell morphology, immunocytochemistry and histology. In contrast the trabecular derived cells displayed a lower differentiation potential. These findings indicate that there are subpopulations within the isolated cell populations allowing all these cell preparations to express adipocyte-, osteoblast-, chondrocyte-, and neuron-like phenotypes.

Chapter 4 - Contrary to long-held dogma, neurogenesis occurs in the adult brain and neural stem cells (NSCs) reside in the adult central nervous system (CNS). This suggests that the CNS has the potential to self-repair. In support of this contention, neurogenesis is modulated in the adult brain, new neuronal cells are generated at sites of degeneration and injuries, after transplantation adult-derived neural progenitor and stem cells adopt the fate of the tissue, and adult stem cells may have the potential to differentiate into other lineages. Niches are microenvironments that control stem cell activity. The therapeutic potential of stem cells, particularly adult NSCs, lies therefore in the interaction between stem cells and their niches. Neurogenic niches have been identified in the adult brain. Unraveling the interactions between NSCs and their niches will contribute to our understanding of the developmental potential of adult stem cells, and to bring adult NSC research to therapy.

Chapter 5 - With the passing of the Human Fertilization and Embryology (Research Purposes) Regulations 2001, the United Kingdom of Great Britain and Northern Ireland became the first country to pass legislation in support of embryonic stem cell research and research on embryos created by somatic cell nuclear transfer. While the UK legal stance and framework on embryo research are well-known and have attracted significant attention from those with an interest in the ethics of embryo research both in the UK and elsewhere, the arguments on the status of the embryo that were expressed by the members of Parliament and the main advisory bodies involved in this legal debate are less well-known. This article examines the entire range of arguments expressed in support of the Regulations. When the Human Fertilization and Embryology (Research Purposes) Regulations 2001 were passed, the UK had already established a legal framework on embryo research. Since this new legal stance must be understood against the background of these earlier developments and discussions on the status of the embryo, this article will also sketch the legal history of embryo research in the UK, documenting in particular how this legal history has been influenced by the Committee of Inquiry into Human Fertilization and Embryology's arguments on the status of the embryo. While the author argues that the validity of all the arguments can be questioned, the authors also argues that, as long as people have irreconcilable values, no case either for or against granting full moral status to the early embryo can be made that will convince everyone. At the same time, by clarifying my own position on the status of the early embryo, The author hopes to throw some light on why the author believes embryonic stem cell research should not be carried out. Therefore, this article will be of interest to all who have an interest in the ethics of embryonic stem cell research.

Chapter 6 - Inducible transgene expression remains an important challenge in embryonic stem cell research. It is desirable because it allows to control the level and timing of gene expression in stem cells. Here the authors describe a tetracycline-regulatable system based on the repression of a modified EF1-α promoter (EF1-αTetO2) by the native tetracycline repressor. EF1-αTetO2 activity was low in tetracycline repressor-expressing cell lines, but was strongly enhanced upon addition of tetracycline. Effective transgene expression was achieved within 24 hours of tetracycline addition; expression levels returned to baseline within 24 hours after tetracycline withdrawal. Graded transgene expression levels were achieved by increasing tetracycline concentrations. Efficient induction was observed in undifferentiated ES cells as well as ES cell-derived neurons. Upon subcutaneous cell implantation in mice, the system allowed in vivo induction of transgene expression. To our understanding, this is the first description of a switch-on expression system allowing transgene induction in implanted ES cells.

Chapter 7 - Mesenchymal stem cells (MSC) can be isolated from virtually all adult and fetal tissues and expanded in vitro. They share their origin in a perivascular niche suggesting a relationship with pericytes. Adult subcutaneous tissue is an abundant source of MSC, which share with bone marrow mesenchymal stem cell (BM-MSC) not only the capacity to differentiate into different mesenchymal lineages, but also an immunomodulatory effect. In view of the capacity to differentiate into different mesodermal tissues, the potential application of adipose tissue-derived MSC (ATMS) in regenerative medicine as an alternative source to BM-MSC has raised great interest. In this chapter, the authors will discuss current data that suggest differences in phenotype, gene expression profile, osteo-chrondogenic potential, and hematopoietic supportive property between AT-MSC and BM-MSC.

Chapter 8 - Dendritic cells (DC) are a heterogenous population of bone marrow-derived cells residing in primary and secondary lymphoid organs. All DC share a common ability to process and present antigen to naïve T cells for the initiation of an immune response. However, they differ in surface markers, migratory patterns, localization and cytokine production. DC had been considered as myeloid cells for some time. However, recent findings have demonstrated that DC can develop not only from myeloid committed progenitors, but also from early lymphoid committed progenitors. The ability of DC to develop from progenitors committed to different lineages correlates with the surface expression of Flt3 receptor. The developmental events downstream from the lymphoid and myeloid committed progenitors as well as the molecular mechanisms required for DC development have been less well characterized. Recent studies have provided evidence for an intra-splenic precursor population that gives rise to the DC subsets *in situ*. The impaired DC development found in mice deficient in certain transcription factors known to be required for hemopoiesis has provided clues to the important role of transcription factors in regulating DC development. The recent finding that steady-state "dendropoiesis" can be recreated *in vitro* using bone marrow progenitors in the presence of Flt3 ligand will facilitate further studies of DC development and function. This review summarizes recent crucial findings on the development of a functional DC system.

Chapter 9 - There has been considerable progress towards the development of a stem cell-based therapy for Hirschsprung's disease (HSCR). This review article discusses the mechanisms regulating enteric nervous system (ENS) stem cell development and the molecular pathogenesis of HSCR in order to elucidate the feasibility of any stem cell therapy. Such a therapy will require a source of stem cells, techniques to isolate and amplify them, and effective methods for the transplantation of the stem cells or their progeny. To date, embryonic and postnatal ENS stem cells have been cultured as neurospheres prior to transplantion into mammalian bowel. However, while there are still many obstacles to overcome, recent observations have demonstrated functional changes in the gut after stem cell transplantation, indicating that a stem cell therapy for HSCR will indeed be possible.

Chapter 10 - The acute respiratory distress syndrome (ARDS) and acute lung injury (ALI) are common diagnoses in intensive care patients who require mechanical ventilation. They are acute in onset and typically fulminant in nature. No medical therapy has been found to improve survival from these disorders despite years of investigation and several clinical trials. Mortality for ARDS continues to approximate 40%. The alveolar epithelium, composed of types I and II cells, is a prime target for damage in ALI/ARDS, as is the pulmonary vasculature. Both the pulmonary epithelium and endothelium are injured diffusely. The development of pulmonary hypertension (PHTN) may be observed as well. Given the lack of available therapies for ALI/ARDS, investigators and clinicians alike are eager to explore new treatment modalities directed at repair of lung damaged in these disorders. As a result, the role of stem cells is receiving increasing attention. This review will focus on the role of stem cells in the repair of lung injury, specifically those studies that focus on the regeneration of lung epithelium and endothelium. Potential therapies for pulmonary hypertension will be highlighted as well. Investigations that have been performed in animal models will be described, and how these may be applicable to human disease.

In: Stem Cell Research Progress
Editor: Prasad S. Koka, pp. 1-17

ISBN: 978-1-60456-308-5
© 2008 Nova Science Publishers, Inc.

Chapter 1

FUNCTIONAL SIGNIFICANCE OF ACTIVE DYE EXCLUSION BY HUMAN HEMATOPOIETIC PROGENITORS AFTER SHORT-TERM CULTURE

Vincent Kindler, Michel A. Duchosal,***
Maya von Planta, Véronique Vallet**,*
*Domizio Suva*** and Caroline Soulas**

*Hematology Service, Geneva University Hospital, Switzerland
**Hematology Service, Lausanne University Hospital, Switzerland
***Orthopedic Surgery Service, Geneva University Hospital, Switzerland

ABSTRACT

Hematopoietic stem cells (HSC) are associated *ex vivo* with a cell subset actively rejecting lipocationic dye such as rhodamine123 (Rho) in the extracellular medium via the multi-drug resistance gene product. It is however not known whether this association remains after culture. We therefore evaluated whether cultures derived from CD34+ human cord blood (CB) cells contained dye-excluding (Rho-low) cells enriched in early HSC. Our investigations showed that after short-term culture with FLT3-ligand, thrombopoietin and stem cell factor, Rho-low cell were still detectable, but with a lower frequency than that of uncultured cells. Rho-low cell frequency varied depending on the stimulus used *in vitro,* and these variations correlated with that of long term-culture initiating cell (LTC-IC) frequencies ($R=0.619$, $p<0.04$), whereas no correlation could be established with late HSC (CFU) frequency. Rho-low cells sorted after 6 days of culture contained 2.8-fold more secondary CFU and 1.8 more LTC-IC compared to Rho-high cells. Though these data suggest that dye exclusion assays could be used as a relevant alternative compared to other *in vitro* assays to identify immature HSC after culture, it remains to be determined whether Rho-low cells derived from cultures retain clinically

---

* Corresponding Author: Dr Vincent Kindler, PhD, Hematology Service, Geneva University Hospital, 25, Micheli-du-Crest, 1211 Geneva 4, Switzerland. tel. ++ 41.22.372.39.42. Fax ++ 41.22.372.72.88. email: vincent.kindler@hcuge.ch

relevant characteristics such as the ability to reconstitute a lethally conditioned host. In this light, we further discuss below the limitations of some *in vitro* and *in vivo* hematopoietic assays, and their relevance as tools to identify HSC that are able rescue conditioned patients.

INTRODUCTION

Clinically relevant hematopoietic stem cells (HSC) are cells able to restore unrestricted and long-term hematopoiesis in a lethally conditioned recipient. High numbers of HSC may be difficult to obtain, and such paucity can preclude their clinical use. For instance, the use of a human cord blood (CB) specimen as a source of transplantable primitive HSC is restricted to patients of less than 40 kilograms and is associated with a prolonged aplasia [1-3]. A better understanding of HSC biology *in vitro* may help devising culture conditions increasing their numbers without a loss of self-renewal [4, 5], and would facilitate therapies involving CB-derived HSC to a larger set of patients. In order to do so, key culture parameters influencing HSC characteristics have to be identified, and specific bioassays aimed at the detection of early HSC are required.

Clonal expansion/differentiation of cells on semi-solid culture was one among the first assays developed to identify hematopoietic precursors [6, 7]. Initially, pluripotent cells, defined as those able to produce colonies consisting in multiple, i.e. granulocytic, erythrocytic, macrophagic and megakaryocytic (CFU-GEMM) lineages [8] were considered to be close, or even identical to, "stem cells". Further sophistication of the clonogenic assays by repicking colonies in secondary cultures demonstrated that only a small fraction of CFU-GEMM was able to generate secondary cultures, suggesting that the CFU-GEMM subset was indeed heterogeneous. Nevertheless such a sequential culture system in semi-solid medium, compared to the single culture system, allows determining the clonogenic frequencies of rather immature hematopoietic progenitors. [9, 10]. Another type of assay, involving the coculture of putative HSC with a stroma, was initially pioneered by Dexter [11-13], and gave birth to the family of long-term culture initiating cell assays, including the long-term culture initiating cell (LTC-IC) assay [14, 15], the cobblestone area forming cell (CAFC) assay [16], and the long-term hematopoietic culture initiating cell (LTHC-IC) assay [17]. In these assays, hematopoietic cells are long-term (6-8 weeks) cocultured on a supporting cellular stroma up to near exhaustion before being tested for their remaining ability to generate colonies on semi-solid medium (LTC-IC), or directly under the supporting stroma (CAFC).

However these *in vitro* assays would hold little significance if not compared with the gold standard of HSC determination that represents the *in vivo* repopulating assay. This assay is based on the cardinal observations of Till and McCulloch demonstrating that a small subset of bone marrow cells reconstituted the hematopoiesis of an entire organism [18]. Original experiments were run in a syngeneic setting in mice. Recently, as a surrogate to human hematopoietic cell engraftment, the assay has been adapted to a xenogeneic setting using the severe combined immunodeficient (SCID)-mouse engrafted with human HSC (SCID-hu) [19-21].

Technically simpler assays, involving the direct detection *ex vivo* of phenotypic markers known to be associated with HSC without requiring cell culture, can also be performed. For instance in humans, bone marrow (BM), CB, or granulocyte colony-stimulating factor (G-

CSF)-mobilized adult blood-derived CD45+CD34+ lineage (lin)- (lin=CD38, CD33, CD42b, CD19, CD2) cells are enriched in HSC [22-24]. Unfortunately the correlation observed between phenotype and HSC function as determined *ex vivo* most often disappears after culture, indicating that phenotypic determination is no more fully valid after *in vitro* manipulation [25].

Finally, HSC can also be identified *ex vivo* by their ability to release in the external medium, via the multi-drug resistant gene complex (MDR), preloaded lipocationic dyes such as rhodamine 123 (Rho) or Hoechst 33342 [26-28]. It is however not clear, from the present literature, how the dye-excluding function is modulated in cultured HSC and whether cells actively excluding Rho are still included within the HSC compartment after culture. In an attempt to answer to these questions, we investigated whether cultures derived from human CD34+ CB cells contained Rho-low cells that were associated with early HSC.

MATERIAL AND METHODS

Purification of CD34+ Cells

Human umbilical cord blood (CB) samples were obtained in accordance with the institutional guidelines of the Ethics Committee of Geneva University Hospital, and CD34+ cells were purified using anti-human CD34 immunomagnetic beads (Dynabeads M-450, Dynal, Oslo, Norway), as described [29]. The mean purity of the CD34+ cells determined by flow cytometry was >90%. These cells will be referred as *"ex vivo"* cells.

Cultures

Liquid cultures: Initial cell density was adjusted to 1-2 x 10^5/mL in 24-well plates. Cells were cultured at 37°C in a 5% CO_2 atmosphere, in Iscove modified Dulbecco's medium (IMDM Gibco BRL, Paisley) complemented with 10% fetal serum (FCS), or in serum-free Cell Gro medium, (Boehringer, Ingelheim, Germany). Both media were supplemented with 10^{-4} M dithiothreitol (DTT, Fluka, Buchs, Switzerland), penicillin and streptomycin (Gibco). Recombinant human thrombopoietin (TPO, 10 U/mL), a mixture of either TPO, FLT3-ligand (FLT3-L, 25 ng/mL), and stem cell factor (SCF, 20 ng/ml) or granulocyte-macrophage colony-stimulating factor (GM-CSF, 20 ng/ml) and interleukin-3 (IL-3 10 ng/ml) were used as growth factors. All were from Peprotech, London, U.K. CD34+ cells were cultured for 4 to 6 days before analysis or cell sorting.

Hematopoietic assays: Clonogenic hematopoietic progenitors (CFU) were assayed by seeding 50 and 100 freshly purified CD34+ cells, or 1000 and 5000 3-week cultured cells in duplicates in 1 mL of 3% methyl cellulose (Fluka Biochemika) containing erythropoietin, G-CSF, GM-CSF, IL-3, SCF, FCS 30%, BSA 0.8%, and 10^{-4} M DTT [30], and colonies were scored 14 days later as described [8]. Secondary CFU were performed by diluting primary CFU cultures 1:10 with Hank's balanced salt solution to wash out the methylcellulose. Cells from primary colonies were counted and seeded at 5000 and 10000 per dish in the above conditions. Colony frequency was assessed after 14 days. LTC-IC were done in replicates in 1 mL cultures, using the MS5 cell line as hematopoietic supporting stroma as described [31]. Cells were cultured for 6 weeks with weekly half-refeeding before counting and assay in methylcellulose as described above.

Flow Cytometry Analysis and Sorting

Immunostaining: The following murine anti-human monoclonal antibodies were utilized: phycoerythrin (PE) labeled anti-CD34 (IgG1, clone HSCA-2) from Becton Dickinson, Mountain View, CA, and anti-CD38 (IgG1, clone HIT2) from Caltag, Burlingame, CA. Mouse IgG1-PE and IgG1-FITC from Dako and Caltag were used as controls.

Rhodamine123 (Rho) exclusion analysis: Rho staining was performed as previously described [32]. Briefly, cells were adjusted to 10^6/mL and incubated for 30 minutes at 37 °C in IMDM containing 2% FCS and 0.1 µg/mL (262 nM) of Rho (Molecular Probes, Eugene, OR). They were then washed twice and incubated for 80 minutes at 37°C with or without 5 mM of the MDR-gene product blocking drug verapamil (Sigma, Saint Louis, MO) to identify cells clearing Rho via an energy-dependent pathway. Rho fluorescence was analyzed in the FL-1 channel. Rho-low cells were defined as cells actively excluding Rho, whose Rho fluorescence detected in absence of verapamil was inferior to the lowest Rho-fluorescence detected in the presence of verapamil. In some instances, cells were then washed with ice-cold PBS containing 3% FCS and 0.02% sodium azide and further stained with PE-or biotin-labeled monoclonal antibodies. Dead cells and debris were excluded using appropriate gates on forward and side scatters, and gating on 7-amino-actinomycin-D low cells [33]. All analyses were run on a FACScalibur (Becton Dickinson).

Cell cycle analysis: The analysis of DNA content was done according to Garvy et al [34] after fixation of the cells in 50% EtOH, RNAse treatment and staining with 50 µg/ml of propidium iodide.

Nucleic acid content of live cells: For nucleic acid detection, cells recovered after the rhodamine123 efflux were incubated for 30 minutes at 4°C in a 1 µM solution of the nucleic acid permeant dye syto 64 (Molecular Probes, Eugene, Oregon). Syto 64 fluorescence was read as FL-4. Cell sorting was performed on a FACSstar+ (Becton Dickinson).

Statistics

Data are presented as mean ± standard deviation. When applicable ($n>3$ per group of measures) statistical significance was determined using the non-parametric Mann-Whitney rank test. Student *t* test is used for samples with $n \geq 8$. Statistical significance was fixed at $p<0.056$ and $p<0.05$ for Mann-Whitney and Student test respectively.

RESULTS

Short-Term Cultures of CD34+ Cells Contain Sizable Frequencies of Rho-Low Cells

Ex vivo Rho-low cells represented 72 ± 9 % of the CD34+ cells (mean $\pm$ SD, *n*= 9, Figure 1A).

After 4 days of culture, Rho-low cells were still detectable but their frequency had decreased: in the presence of early acting growth factors such as TPO, or such as the combination of FLT3-L, TPO, and SCF (FTS), Rho-lo cell frequency represented respectively $20 \pm 11\%$ and 15 ± 8 % of the total in cultures. This value dropped to $4.8 \pm 2.9\%$ when cells were cultured with GM-CSF and IL-3 (*n*=4 for each type of cultures, Figure 1B, left panel). Analysis of total nucleic acid content on live cells showed that Rho-low cell were evenly distributed between the cell fractions with a low content (cells not activated, syto 64^{low}) and a high content (activated cells, syto 64^{high}) of nucleic acid (Figure 1C), suggesting that the decrease in Rho-lo cell frequency was not produced by the preferential sequestration of Rho in the mitochondria of cells activated by hematopoietic growth factors [35, 36]. Thus these data demonstrated that Rho-low cell frequency had decreased in cell suspensions derived from CD34+ cells after culture, and such decrease was the greatest in the cultures performed with factors (*i e* GM-CSF plus IL-3) exhibiting a poor ability to rescue stem cells compared to TPO or FTS.

Relationship between Immature Hematopoietic Progenitor (LTC-IC) and Rho-Low Cell Frequencies after Culture

Cultures derived from CD34+ cord blood cells, obtained after incubation with either TPO alone, or FTS were also assessed for their content in early and late hematopoietic progenitors using the LTC-IC and the CFU-C assays respectively. These analyses showed that LTC-IC frequencies, as reported above for Rho-low cell frequencies, had decreased compared to *ex vivo* values after 4 days of culture, irrespective of the stimulus used (Figure 2A, left panel). By contrast CFU-C frequencies seemed unaffected by the *in vitro* manipulations (Figure 2A right panel).

Rho-low cell frequencies obtained from *ex vivo* CD34+ cell suspensions, or from such suspensions after culture with TPO in absence or presence of serum were plotted against the LTC-IC or the CFU-C frequencies obtained from the corresponding conditions. A linear correlation could be established between Rho-low cell and LTC-IC frequencies (*r*=0.619, *p*=0.04, Figure 2B. left panel), whereas no correlation was identified between Rho-low cell and CFU-C frequencies (Figure 2B, right panel). Altogether these data suggest that Rho-low cells derived from short-term cultures of CD34+ cells are still associated with early HSC.

Cell Cycle and Proliferative Characteristics of Rho-Low and Rho-High Cell Sorted from Culture

To further assess the specific characteristics of Rho-low and Rho high cells surviving *in vitro*, we proceeded to cell sorting according to dye exclusion after a 6 days of culture with FTS in presence of serum. At that time the total cell number was 14 ± 4.5-fold the input value, the Rho-low cell frequency had decreased by half; consequently the absolute number of Rho-low cells had increased by 6.9 ± 2-fold. On the whole we sorted 10 cell suspensions according to the gates depicted on Figure 3A, and immediate reanalysis showed a clear segregation of both cell fractions (Figure 3A, lower panels). Rho-low and Rho-high fractions from 4 cultures were analyzed for cell cycle immediately after sorting. Cells in S-G2M phases represented $30 \pm 2.9\%$ and $36 \pm 3\%$ in Rho-low and Rho-high fractions respectively (Figure 3B). Six experiments were recultured on a short-term basis (5 days) after sort and reanalyzed for rhodamine123 exclusion. These showed without exception that cells initially sorted as Rho-low contained a relevant proportion of cells that conserved their ability to reject Rho very efficiently (Figure 3C).

By contrast cell suspensions sorted as Rho-high were essentially devoid of Rho-low, verapamil-sensitive cells. This was not caused by the persistent sequestration of Rho in the cells originally sorted as Rho-high because such cells exhibited a very low basal autofluorescence if not re-exposed to Rho (see Figure 3C, control).

Rho-low and Rho-high sorted cells were also incubated on a long term basis with repetitive culture medium replacements. Both fractions were indeed able to sustain long-term proliferation, but with different kinetics: the mean cell-doubling time was significantly shorter (5.5 days *vs.* 9.4 days, $p<0.016$), and the global cell yield after 38 days of culture was 4.6-fold higher for the Rho-low cells compared to the Rho-high cells (Figure 3D). Thus collectively these observations showed that the Rho-low cells identified after 6 days of culture in FLT3-L, TPO, and SCF contained by contrast with the Rho-low cells isolated *ex vivo* [37, 38] a relevant fraction of cells that was in cycle. Moreover they harbored a pool of long-lasting Rho-low cells that was not present in the Rho-high sorted cell and generated, on a long range, a larger progeny compared to the Rho-high cells sorted from the cultures on day 6. All these observations argue in favor of enrichment in immature hematopoietic progenitors in the Rho-low, compared to the Rho high sorted fraction obtained after culture.

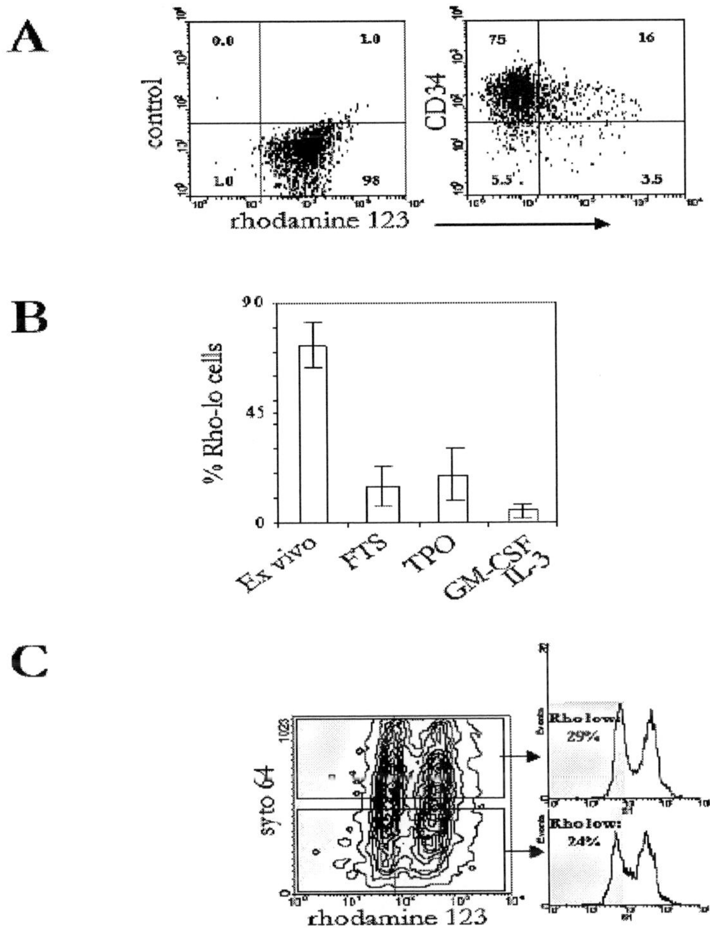

Figure 1. Rho exclusion characteristics of CD34+ CB cells before and after culture. (A); Bead-purified CD34+ CB cells were analyzed before culture by flow cytometry for CD34 expression and Rho exclusion. Left panel: controls; Rho fluorescence after desaturation in presence of verapamil on x-axis, and isotype control on y-axis. Right panel: Rho fluorescence after desaturation in absence of verapamil on x-axis, and staining with anti-CD34 mAb on y-axis. The percentages of live cells within each quadrant are indicated. These data are representative of 9 independent experiments. (B); Mean percentage of Rho-low cells (as defined in Materials and Methods) before and after 4 days of culture of CD34+ cells with FLT3-L, TPO, and SCF (FTS), TPO alone or GM-CSF plus IL-3. $n=4$, for all conditions except for GM-CSF+IL-3 cultures where $n=3$, SD are shown. (C) Nucleic acid content of cells derived from CD34+CB cells after 4 days of culture. Syto 64 fluorescence, which is representative of nucleic acid content, is on y-axis while Rho-123 fluorescence is on the x-axis. The shaded area correspond to Rho-low cells as determined using verapamil. Boxes determine the gates used to build up the profiles of Rho fluorescence in the nucleic acid-high and nucleic acid-low fractions shown on the right. These data are representative of 3 independent experiments.

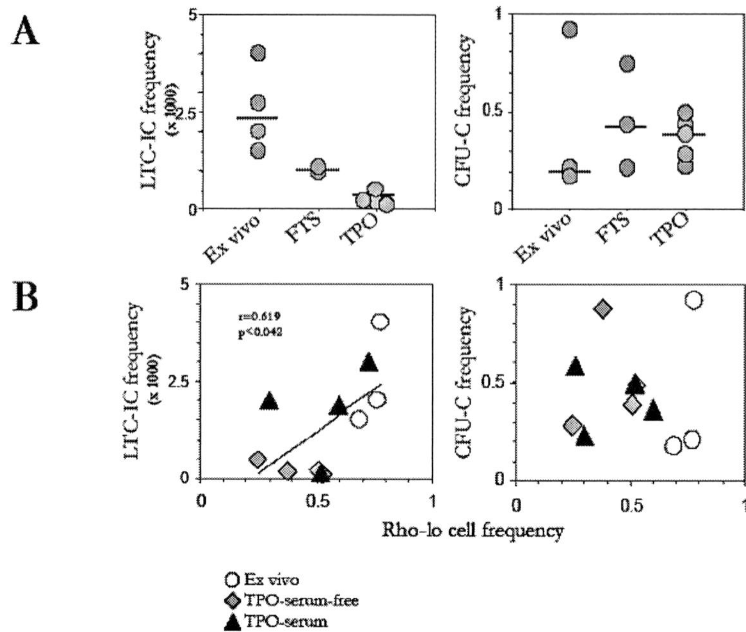

Figure 2. Hematopoietic content of *ex vivo* and cultured cells. (A) LTC-IC (left) and CFU (right) absolute frequency (as computed relative to 1) analyzed either immediately after CD34+ cell purification or after 4 days of culture with TPO or FTS in serum free conditions. Each circle represents a single independent experiment and the horizontal bars represent the medians. (B) Relationship between Rho-low cell and LTC-IC frequency (left) and Rho-low cell and CFU frequency (right) under different experimental conditions. Open circle represent *ex vivo* cells, gray diamonds 4 days of culture with TPO in serum-free conditions, and the solid triangles, 4 days of culture with TPO in presence of 10% FCS. Each point used to build up the plots is one independent experiment. Rho-low cell frequencies are on the x-axis, and LTC-IC and CFU frequencies are on the y-axis. Frequencies are represented as absolute values. A linear correlation could be established between Rho-low cell and LTC-IC frequencies only.

Early Hematopoietic Progenitor Content in Rho-Low and Rho-High Cells Obtained from Sorted Cultures

Rho-low and Rho-high cell suspensions obtained after cell sorting of primary cultures as above were evaluated for their content in various types of HSC. Mean CFU frequencies measured immediately after sorting were identical in both fractions (n=3, Figure 4), indicating that cultured Rho-low and Rho-high cells had a similar content in late hematopoietic progenitors. By contrast, immature HSC frequencies were different between the 2 fractions. Secondary CFU (i.e., CFU obtained after replating of the primary CFU), and LTC-IC (i.e., CFU obtained after clonogenic culture of cells derived from a 6 week-liquid culture on the MS-5 stroma) frequencies averaged respectively $2.7 \times 10^{-3} \pm 5.7 \times 10^{-4}$ and $3.0 \times 10^{-3} \pm 2.8 \times 10^{-4}$ in Rho-low cell suspensions, and $9.5 \times 10^{-4} \pm 2.1 \times 10^{-4}$ and $1.6 \times 10^{-3} \pm 1.8 \times 10^{-4}$ in Rho-high cell suspensions (mean value of duplicate $\pm$ SD of one experiment representative of 2 for secondary CFU, and of a single experiment for LTC-IC, see Figure 4).

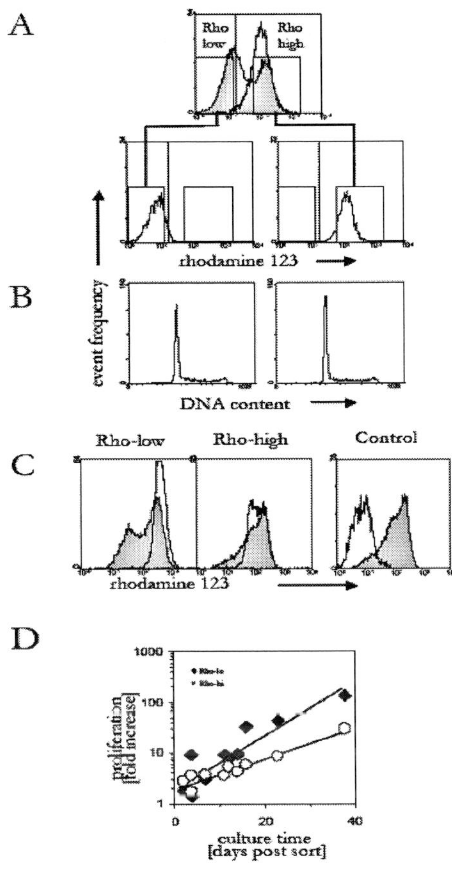

Figure 3. Cell cycling and proliferative properties of Rho-low and Rho-high cells sorted from cultures. CD34+ cells derived from CB were cultured for 6 days with FLT3-L, TPO, and SCF in presence of serum and were incubated with rhodamine123. Cells were then desaturated with or without verapamil as in Figure 1, and sorted as Rho-low and Rho-high cells. (A); sorting gates and reanalysis. (B); DNA content of Rho-low and Rho-high cells immediately after sorting. Cell cycle analysis was performed after RNAse treatment on ethanol-permeabilized cells using propidium iodide to stain DNA. These data are representative of 4 independent experiments. Numbers represent the % of cells in SG$_2$M phases. (C); Rho-exclusion properties of sorted Rho-low and Rho-high cells after a secondary 5 day-culture in FTS. White and gray profiles represent Rho-123 fluorescence with, respectively without, verapamil during desaturation. Control: the control profiles are Rho-high sorted cells that have been recultured and either analyzed as such (white profile) or after labeling with Rho and desaturation in presence of verapamil. This is to show that they were devoid of residual rhodamine123 derived from the cell sorting procedure. These data are from one experiment representative of six. (D); long-term proliferation of Rho-low and Rho-high cells in secondary cultures after sorting. According to double regression analysis, Rho-low and Rho-high cell doubling time were significantly different ($p<0.016$). These data are the pool of 3 independent experiments.

Thus the secondary CFU and the LTC-IC frequencies as read directly at the end point of these assays were respectively 2.8- and 1.8-fold higher in Rho-low compared to the Rho-high sorted cell fraction. However since cells proliferate in primary CFU-C and in liquid LTC-IC cultures (not shown), and since only a fraction of these cells are introduced in secondary clonogenic assays required for both approaches (see the Materials and Methods section), the rough frequencies recorded for these cell suspensions as depicted above underestimate the overall frequency of early hematopoietic progenitors initially seeded in the hematopoietic assays. To get a more accurate estimation of such frequency one has to combine the cell proliferation observed during primary CFU and LTC-IC cultures with the rough frequencies. Introducing these values in our computations allows to state that sorted Rho-low cells recovered after culture with FTS produced 5.6 secondary CFU and 0.14 LTC-IC per cell, whereas Rho-high cells sorted out from the same source produced 1.2 secondary CFU and 1.6 x 10^{-3} LTC-IC per cell.

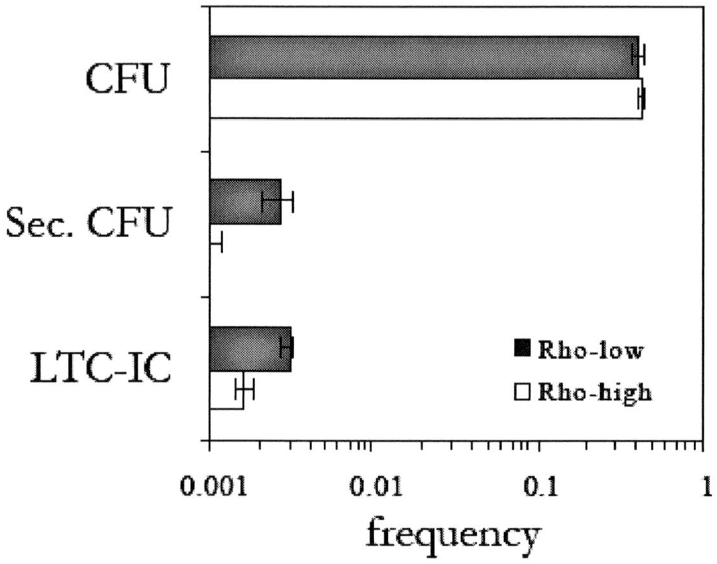

Figure 4. HPC content of Rho-low and Rho-high cells sorted after culture. CD34+ cells derived from CB were cultured for 6 days with FLT3-L, TPO, and SCF in presence of serum, were incubated with rhodamine123 and sorted as Rho-low and Rho-high cells as above and subjected to *in vitro* assessment of hematopoietic progenitor frequency. Gray symbols refer to Rho-low cells, and white symbols to Rho-high cells. *CFU:* CFU frequency; *sec. CFU:* secondary CFU frequency; *LTC-IC:* LTC-IC frequency. These data are the mean ± SD of duplicates from one single experiment representative of 3 for CFU, and 2 for secondary CFU. LTC-IC frequency on sorted fractions has been assayed once. Frequencies are expressed as absolute values (expressed relative to 1).

DISCUSSION

Our study showed that cells derived from short-term cultures of CD34+ CB cells contained cells able to exclude rhodamine123 via the MDR, in a verapamil-sensitive manner, and that the frequency of Rho-low cells recovered from the cultures had decreased compared to *ex vivo* values. Rho-low cell maintenance in cultures seemed to be favored by growth factors such as TPO, or TPO plus FLT3-L and SCF, which are known to produce modest HSC differentiation *in vitro* [39], but not by GM-CSF and IL-3 that induced a more extensive decrease in Rho-low cells.

The variation of Rho-low frequencies correlated with that of early, but not late HSC. This suggested that early HSC were still included in the Rho-low subset recovered after *in vitro* manipulations. Moreover sorting cells from primary cultures further showed that Rho-low cells were enriched, compared to their Rho-high counterpart, in early HSC as identified by *in vitro* hematopoietic bioassays.

In our cultures Rho-low cells were detected with similar frequencies in nucleic acid^{low} and nucleic acid^{high} cell fractions, indicating that both non-activated and activated cells excluded Rho in a verapamil-sensitive process. In young mice rhodamine123 can be sequestrated very efficiently within activated mitochondria and be, in this compartment, out of reach of the MDR activity. Such a situation can lead to the generation of verapamil-resistant, Rho-high cells that may nevertheless be pluripotent ancestral stem cells [35, 36]. Our above observation combined with our analyses of cell cycle showing that both Rho-low and Rho-high cells contained cells in S-G2M suggested that the eventual labeling of mitochondria by Rho in the activated cells was not sufficient to occult MDR-driven Rho-efflux in short-term cultures of human cells.

By contrast to *ex vivo* [37, 38], Rho-low cells derived from cultures contained a significant proportion of cycling cells. The total number of Rho low cells increased relative to the *ex vivo* values, strongly supporting that cycling Rho-low cells survive long enough to complete several cell cycles and produce an extensive progeny while keeping their ability to exclude Rho via the MDR. However this property appears transient because primary cultured cells subsequently sorted as a homogenous Rho-low cell suspension and grown in secondary cultures generated a large proportion of Rho-high cells. Thus the ability to exclude rhodamine123 is short-lived within the cultures, and cells will eventually differentiate irreversibly into the Rho-high phenotype. Indeed, cells sorted as Rho-high after primary cultures did not recover their ability to exclude Rho after secondary cultures. These Rho-high cells may cycle for a certain time but, as suggested by our sorting experiments, they exhibit a restricted proliferation compared to Rho-low cells, and may rapidly differentiate into Rho-high non-cycling cells.

It is difficult to establish whether or not Rho-low cells recovered from the culture are pluripotent hematopoietic stem cells. The correlation between the Rho-low cell- and LTC-IC frequencies favors this contention, but the later frequencies were at least 100-fold lower than that of the Rho-low cells. This suggests that only a small fraction of the Rho-low cells from primary cultures sustained the long- term survival of a hematopoietic progeny over a competent stroma. Thus it is likely that most of the Rho-low cells rescued in our cultures are indeed pluripotent progenitors, but with a limited self-renewal potential. In agreement with this hypothesis, our preliminary *in vivo* repopulating experiments in NOD/SCID mice using Rho-low cell suspensions sorted from short term-culture suggest that the proportion of repopulating human HSC at 6 weeks is very small in these suspensions (MAD and VK, unpublished).

True, self-renewing stem cells are believed to be essentially in G0 and tightly secured in this cycle state by their interaction with their specific environment, identified as the "niche" [40]. The niche provides all the molecular mediators required to warrant the maintenance and survival of stem cells [for review, see [41, 42]]. Moreover, the hematopoietic niche seems to be subdivided in two distinct parts, the vascular unit and the hypoxic unit, which both fulfilling specific tasks. In steady state conditions (tissue homeostasis), the vascular unit sustains asymmetric mitoses, giving birth to one ancestral stem cell identical to its mother cell and one transit stem cell that will participate to tissue maintenance and regeneration. The hypoxic unit furnishes all the required factors to protect stem cell integrity [43] and preserve its quiescence. Stem cells from this compartment are solicited to enter mitosis only when tissue homeostasis is disrupted abruptly and the replenishment of the vascular unit is required. Most likely, this occurs after an extended injury has prompted the massive emigration of the vascular niche content toward the injured site to restore tissue function.

In order to ensure stem cell self-renewal, the HSC niche has developed a complex architecture comprising several cell types and a wide array of molecular factors distributed in a very precise tri-dimensional structure. HSC have to be localized in the right places within this structure to maintain their state and/or respond to tissue needs. We are just beginning to understand the complexity of this structure and our current *in vitro* systems poorly mimic the *in vivo* situation. Indeed, cultures set up with soluble cytokines only are obviously devoid of specific cell-cell interactions required for HSC to self-renew. Additionally, cocultures of HSC with stroma (often under xenogeneic conditions), though being closer to an *in vivo* situation compared to the previous systems, certainly lack as well several components crucial for HSC self-renewal: in these cocultures, monotypic cells from established lines are generally used, and these may not be sufficiently diverse in terms of surface molecule presentation to provide all the interactions occurring between the HSC and the various constituents of its niche *in vivo*. This hypothesis is supported by the absence of lymphoid progenitor generation and the rapid disappearance of erythroids in classical LTC-IC [44]. Moreover, most cultures are performed with a limited control of the oxygen tension, which consequently will stabilize close to the atmospheric value ($\approx$20%) in the incubators. This will in turn establish in culture media O_2 titers excessively elevated compared to those generally observed in tissues *in vivo* [45], and especially to these prevailing in the hypoxic niche [43]. Interestingly, such elevated oxygen concentrations generated *in vitro* may favor stem cell differentiation [46].

In the absence of a proper niche, cells tend to differentiate and stem cells are lost [40, 47]. In lethally conditioned hosts total body irradiation and chemotherapy kill the HSC and empty the niches. Upon exogenous HSC infusion niches will be recolonized. The extend of hematopoiesis restoration will eventually correlate with the number of stem cell initially injected, as demonstrated by competitive repopulating assays in mice [48]. However, hematopoietic reconstitution may not always be integral. In some instances, a fraction of the niches also may be irreversibly destroyed by the conditioning regimen, thus leading to a partial recovery of the hematopoiesis process compared to pre-regimen values, irrespective of the "quality" of the HSC injected. Thus niches are the key factors limiting the number of self-renewing, functional stem cells. A similar situation occurs when one tries to amplify stem cells *in vitro*. Thanks to cell sorting, we can now identify, purify and seed highly potent HSC in culture systems, but most often these cells differentiate. This is not linked to their stemness *per see* but rather to the absence of the appropriate microenvironment. An *in vitro* culture system will secure the stem cells seeded, or eventually amplify them only if it furnishes proper niches. Even so, the stem cell numbers generated *in vitro* are not expected to exceed that of the niches provided by the system. This evidence emphasizes the need to decipher all the components of the niche in order to be able to reconstruct it *in vitro*. This is the prerequisite to any attempt aimed at the preservation, or perhaps the amplification of stem cells, a process that *in vitro* as well as *in vivo* will persist only until all the (engineered) niches are replete [49, 50].

Since the description of the CFU-S assay by McCulloch and Till [51] the *in vivo* repopulating abilities of HSC have been widely explored in mice and several cell subsets exhibiting differential repopulating characteristics have been identified (see [42, 44]). These are related to their states of differentiation (i.e. a cell close to final differentiation will not be able to generate as many blood cell lineages as ancestral stem cells would), and also to their cycle status at the time of injection into the host. Indeed, stem cell progression through the cell cycle is associated with a decrease in self-renewal abilities upon transplantation in a compromised host, but such a process is reversible upon cycle arrest [52-54].

Approximately 20-40% of total bone marrow cells injected i.v. in syngeneic animals migrate within a few hours to the BM, which by definition contains the fittest niches to host the colonizing HSC. When human cells are injected in NOD-SCID mouse only 1% of such human medullar cells xenotransplanted migrates to NOD-SCID mouse BM [55, 56], suggesting that a relevant fraction of progenitor cells do not find their niche and may be lost during the xenograft process. NOD-SCID mouse bone marrow niches are highly diverse with regard to their cell composition, which contrasts to the monotypic cell distribution of *in vitro* coculture experiments with stromal cell lines, an observation that would strengthen the more physiological diversity of hematopoietic supporting stroma in this *in vivo* model over that of classical *in vitro* assays. It remains that the murine niche offered to the human cells may not be optimized to assist stem cell self-renewal because many murine hematopoietic factors cannot stimulate human cells (*i.e.* IL-3, GM-CSF, IL-6), or do so at high (still physiological?) concentrations (Epo, SCF, IL-7) compared to their human counterparts. These observations confirm that the murine hematopoietic environment allows the homing and survival of human HSC *in vivo*, but suggest that the frequency of human HSC capable to restore the human hematopoietic system in autologous or allogeneic hosts is higher than that of the NOD-SCID repopulating cells. This also indicates, if the model is able to distinguish qualitative patterns of human hematopoiesis *in vivo* [11], that it has probably a low sensitivity when compared to clinical settings with regards to quantitative determination of hematopoietic expansion.

In summary this work illustrated that hematopoietic stem cells, though easily identifiable as a concept, and certainly functional as demonstrated by the inverse correlation linking the number of human CD34+ cells injected into aplastic patients with the delay of hematopoietic reconstitution [57] are extremely versatile. Their biological characteristics are not constant, are highly related to their environment, and the assays used to quantify them, especially those concerning human HSC, do not provide a suitable niche that allows the full safeguard of stem cell integrity. Hopefully our understanding of the soluble and membrane-bound mediators, as well as of the tissue organization involved in HSC survival have significantly increased during these last few years and will help to develop culture systems and bioassays that may open up new avenues in cell therapies involving HSC. In this line, the assessment of dye excluding cells in cultures derived from hematopoietic cells, though certainly not identifying exclusively HSC, appears as a valid approach as a first and rapid screening procedure to allow identifying culture conditions that favor hematopoietic stem cell survival.

ACKNOWLEDGMENTS

We thank Nathalie Gardiol, Dominique Aubry, and Nicolette Brouwers for excellent technical assistance. This work was supported in part by the Novartis Foundation for Medico-Biological Science Research and by the Swiss National Science Foundation to MAD, and by the Foundation Dr Henri Dubois-Ferrière-Dinu Lipatti to VK. Grateful thanks also to Pr Bernard Chapuis for his faultless support.

REFERENCES

[1]　Gluckman E. Current status of umbilical cord blood hematopoietic stem cell transplantation. *Exp. Hematol.* (2000) 28 (11): 1197-1205.

[2]　Kurtzberg J, Laughlin M, Graham ML, et al. Placental blood as a source of hematopoietic stem cells for transplantation into unrelated recipients. *N. Engl. J. Med.* (1996) 335 (3): 157-166.

[3]　Rubinstein P, Carrier C, Scaradavou A, et al. Outcomes among 562 recipients of placental-blood transplants from unrelated donors. *N. Engl. J. Med.* (1998) 339 (22): 1565-1577.

[4]　Hoffman R. Progress in the development of systems for in vitro expansion of human hematopoietic stem cells. *Curr. Opin. Hematol.* (1999) 6 (3): 184-191.

[5]　Morrison SJ, Shah NM, Anderson DJ. Regulatory mechanisms in stem cell biology. *Cell* (1997) 88 (3): 287-298.

[6]　Iscove NN, Senn JS, Till JE, McCulloch EA. Colony formation by normal and leukemic human marrow cells in culture: effect of conditioned medium from human leukocytes. *Blood* (1971) 37 (1): 1-5.

[7]　Stanley ER, Metcalf D, Maritz JS, Yeo GF. Standardized bioassay for bone marrow colony stimulating factor in human urine: levels in normal man. *J. Lab. Clin. Med.* (1972) 79 (4): 657-668.

[8]　Fauser AA, Messner HA. Fetal hemoglobin in mixed hemopoietic colonies (CFU-GEMM), erythroid bursts (BFU-E) and erythroid colonies (CFU-E): assessment by radioimmune assay and immunofluorescence. *Blood* (1979) 54 (6): 1384-1394.

[9]　Carow CE, Hangoc G, Broxmeyer HE. Human multipotential progenitor cells (CFU-GEMM) have extensive replating capacity for secondary CFU-GEMM: an effect enhanced by cord blood plasma. *Blood* (1993) 81 (4): 942-949.

[10]　Pierelli L, Scambia G, Bonanno G, et al. CD34+/CD105+ cells are enriched in primitive circulating progenitors residing in the G0 phase of the cell cycle and contain all bone marrow and cord blood CD34+/CD38low/- precursors. *Br. J. Haematol.* (2000) 108 (3): 610-620.

[11]　Dexter TM, Allen TD, Lajtha LG. Conditions controlling the proliferation of haemopoietic stem cells in vitro. *J. Cell Physiol.* (1977) 91 (3): 335-344.

[12]　Dexter TM. Haemopoiesis in long-term bone marrow cultures. A review. *Acta Haematol.* (1979) 62 (5-6): 299-305.

[13]　Dexter TM, Wright EG, Krizsa F, Lajtha LG. Regulation of haemopoietic stem cell proliferation in long term bone marrow cultures. *Biomedicine* (1977) 27 (9-10): 344-349.

[14]　Sutherland HJ, Eaves CJ, Eaves AC, Dragowska W, Lansdorp PM. Characterization and partial purification of human marrow cells capable of initiating long-term hematopoiesis in vitro. *Blood* (1989) 74 (5): 1563-1570.

[15] Eaves CJ, Cashman JD, Sutherland HJ, et al. Molecular analysis of primitive hematopoietic cell proliferation control mechanisms. *Ann. NY Acad. Sci.* (1991) 628: 298-306.

[16] Ploemacher RE, van der Sluijs JP, Voerman JS, Brons NH. An in vitro limiting-dilution assay of long-term repopulating hematopoietic stem cells in the mouse. *Blood* (1989) 74 (8): 2755-2763.

[17] Traycoff CM, Kosak ST, Grigsby S, Srour EF. Evaluation of ex vivo expansion potential of cord blood and bone marrow hematopoietic progenitor cells using cell tracking and limiting dilution analysis. *Blood* (1995) 85 (8): 2059-2068.

[18] Till JE, McCulloch EA. A direct measurement of the radiation sensitivity of normal mouse bone marrow cells. *Radiat. Res.* (1961) 14: 213-222.

[19] Kamel-Reid S, Dick JE. Engraftment of immune-deficient mice with human hematopoietic stem cells. *Science* (1988) 242 (4886): 1706-1709.

[20] McCune JM, Namikawa R, Kaneshima H, Shultz LD, Lieberman M, Weissman IL. The SCID-hu mouse: murine model for the analysis of human hematolymphoid differentiation and function. *Science* (1988) 241 (4873): 1632-1639.

[21] Pflumio F, Izac B, Katz A, Shultz LD, Vainchenker W, Coulombel L. Phenotype and function of human hematopoietic cells engrafting immune-deficient CB17-severe combined immunodeficiency mice and nonobese diabetic-severe combined immunodeficiency mice after transplantation of human cord blood mononuclear cells. *Blood* (1996) 88 (10): 3731-3740.

[22] Bensinger WI, Buckner CD, Shannon-Dorcy K, et al. Transplantation of allogeneic CD34+ peripheral blood stem cells in patients with advanced hematologic malignancy. *Blood* (1996) 88 (11): 4132-4138.

[23] Hassan HT, Zeller W, Stockschlader M, Kruger W, Hoffknecht MM, Zander AR. Comparison between bone marrow and G-CSF-mobilized peripheral blood allografts undergoing clinical scale CD34+ cell selection. *Stem Cells* (1996) 14 (4): 419-429.

[24] Terstappen LW, Huang S, Safford M, Lansdorp PM, Loken MR. Sequential generations of hematopoietic colonies derived from single nonlineage-committed CD34+CD38- progenitor cells. *Blood* (1991) 77 (6): 1218-1227.

[25] Dorrell C, Gan OI, Pereira DS, Hawley RG, Dick JE. Expansion of human cord blood CD34(+)CD38(-) cells in ex vivo culture during retroviral transduction without a corresponding increase in SCID repopulating cell (SRC) frequency: dissociation of SRC phenotype and function. *Blood* (2000) 95 (1): 102-110.

[26] Chaudhary PM, Roninson IB. Expression and activity of P-glycoprotein, a multidrug efflux pump, in human hematopoietic stem cells. *Cell* (1991) 66 (1): 85-94.

[27] Feuring-Buske M, Hogge DE. Hoechst 33342 efflux identifies a subpopulation of cytogenetically normal CD34(+)CD38(-) progenitor cells from patients with acute myeloid leukemia. *Blood* (2001) 97 (12): 3882-3889.

[28] Udomsakdi C, Eaves CJ, Sutherland HJ, Lansdorp PM. Separation of functionally distinct subpopulations of primitive human hematopoietic cells using rhodamine-123. *Exp. Hematol.* (1991) 19 (5): 338-342.

[29] Arrighi JF, Hauser C, Chapuis B, Zubler RH, Kindler V. Long-term culture of human CD34(+) progenitors with FLT3-ligand, thrombopoietin, and stem cell factor induces extensive amplification of a CD34(-)CD14(-) and a CD34(-)CD14(+) dendritic cell precursor. *Blood* (1999) 93 (7): 2244-2252.

[30] Arrighi JF, Zubler R, Hauser C, et al. CD34(+) cord blood cells expressing cutaneous lymphocyte-associated antigen are enriched in granulocyte-macrophage progenitors and support extensive amplification of dendritic cell progenitors. *Exp. Hematol.* (2001) 29 (8): 1029-1037.

[31] Berardi AC, Meffre E, Pflumio F, et al. Individual CD34+CD38lowCD19-CD10- progenitor cells from human cord blood generate B lymphocytes and granulocytes. *Blood* (1997) 89 (10): 3554-3564.

[32] Uchida N, Combs J, Chen S, Zanjani E, Hoffman R, Tsukamoto A. Primitive human hematopoietic cells displaying differential efflux of the rhodamine 123 dye have distinct biological activities. *Blood* (1996) 88 (4): 1297-1305.

[33] Schmid I, Uittenbogaart CH, Keld B, Giorgi JV. A rapid method for measuring apoptosis and dual-color immunofluorescence by single laser flow cytometry. *J. Immunol. Methods* (1994) 170 (2): 145-157.

[34] Garvy BA, Telford WG, King LE, Fraker PJ. Glucocorticoids and irradiation-induced apoptosis in normal murine bone marrow B-lineage lymphocytes as determined by flow cytometry. *Immunology* (1993) 79 (2): 270-277.

[35] Chen LB, Summerhayes IC, Johnson LV, Walsh ML, Bernal SD, Lampidis TJ. Probing mitochondria in living cells with rhodamine 123. *Cold Spring Harb. Symp. Quant. Biol.* (1982) 46 Pt 1: 141-155.

[36] Kim M, Cooper DD, Hayes SF, Spangrude GJ. Rhodamine-123 staining in hematopoietic stem cells of young mice indicates mitochondrial activation rather than dye efflux. *Blood* (1998) 91 (11): 4106-4117.

[37] Leitner A, Strobl H, Fischmeister G, et al. Lack of DNA synthesis among CD34+ cells in cord blood and in cytokine-mobilized blood. *Br. J. Haematol.* (1996) 92 (2): 255-262.

[38] Liu H, Verfaillie CM. Myeloid-lymphoid initiating cells (ML-IC) are highly enriched in the rhodamine-c-kit(+)CD33(-)CD38(-) fraction of umbilical cord CD34(+) cells. *Exp. Hematol.* (2002) 30 (6): 582-589.

[39] Luens KM, Travis MA, Chen BP, Hill BL, Scollay R, Murray LJ. Thrombopoietin, kit ligand, and flk2/flt3 ligand together induce increased numbers of primitive hematopoietic progenitors from human CD34+Thy-1+Lin- cells with preserved ability to engraft SCID-hu bone. *Blood* (1998) 91 (4): 1206-1215.

[40] Schofield R. The relationship between the spleen colony-forming cell and the haemopoietic stem cell. *Blood Cells* (1978) 4 (1-2): 7-25.

[41] Kindler V. Postnatal stem cell survival: does the niche, a rare harbor where to resist the ebb tide of differentiation, also provide lineage-specific instructions? *J. Leukoc. Biol.* (2005) 78 (4): 836-844.

[42] Wilson A, Trumpp A. Bone-marrow haematopoietic-stem-cell niches. *Nat. Rev. Immunol.* (2006) 6 (2): 93-106.

[43] Chow DC, Wenning LA, Miller WM, Papoutsakis ET. Modeling pO(2) distributions in the bone marrow hematopoietic compartment. II. Modified Kroghian models. *Biophys. J.* (2001) 81 (2): 685-696.

[44] Coulombel L. Identification of hematopoietic stem/progenitor cells: strength and drawbacks of functional assays. *Oncogene* (2004) 23 (43): 7210-7222.

[45] Shafer JA, Davis AR, Olmsted-Davis EA. Control of tissue regeneration through oxygen concentrations. *JOSC* (2006) 1 (1): 87-95.

[46] Mostafa SS, Papoutsakis ET, Miller WM. Oxygen tension modulates the expression of cytokine receptors, transcription factors, and lineage-specific markers in cultured human megakaryocytes. *Exp. Hematol.* (2001) 29 (7): 873-883.

[47] Potten CS, Loeffler M. Stem cells: attributes, cycles, spirals, pitfalls and uncertainties. Lessons for and from the crypt. *Development* (1990) 110 (4): 1001-1020.

[48] Szilvassy SJ, Humphries RK, Lansdorp PM, Eaves AC, Eaves CJ. Quantitative assay for totipotent reconstituting hematopoietic stem cells by a competitive repopulation strategy. *Proc. Natl. Acad. Sci. USA* (1990) 87 (22): 8736-8740.

[49] Zhang J, Niu C, Ye L, et al. Identification of the haematopoietic stem cell niche and control of the niche size. *Nature* (2003) 425 (6960): 836-841.

[50] Calvi LM, Adams GB, Weibrecht KW, et al. Osteoblastic cells regulate the haematopoietic stem cell niche. *Nature* (2003) 425 (6960): 841-846.

[51] McCulloch EA, Till JE. Proliferation Of Hemopoietic Colony-Forming Cells Transplanted Into Irradiated Mice. *Radiat. Res.* (1964) 22: 383-397.

[52] Habibian HK, Peters SO, Hsieh CC, et al. The fluctuating phenotype of the lymphohematopoietic stem cell with cell cycle transit. *J. Exp. Med.* (1998) 188 (2): 393-398.

[53] Colvin GA, Lambert JF, Abedi M, et al. Differentiation hotspots: the deterioration of hierarchy and stochasm. *Blood Cells Mol. Dis.* (2004) 32 (1): 34-41.

[54] Zipori D. The stem state: plasticity is essential, whereas self-renewal and hierarchy are optional. *Stem Cells* (2005) 23 (6): 719-726.

[55] Cashman JD, Lapidot T, Wang JC, et al. Kinetic evidence of the regeneration of multilineage hematopoiesis from primitive cells in normal human bone marrow transplanted into immunodeficient mice. *Blood* (1997) 89 (12): 4307-4316.

[56] Cashman J, Bockhold K, Hogge DE, Eaves AC, Eaves CJ. Sustained proliferation, multi-lineage differentiation and maintenance of primitive human haemopoietic cells in NOD/SCID mice transplanted with human cord blood. *Br. J. Haematol.* (1997) 98 (4): 1026-1036.

[57] To LB, Haylock DN, Simmons PJ, Juttner CA. The biology and clinical uses of blood stem cells. *Blood* (1997) 89 (7): 2233-2258.

In: Stem Cell Research Progress
Editor: Prasad S. Koka, pp. 19-30

ISBN: 978-1-60456-308-5
© 2008 Nova Science Publishers, Inc.

Chapter 2

THE EFFECTS OF EX VIVO EXPANSION ON THE RHEOLOGY OF UMBILICAL CORD BLOOD-DERIVED CD34+ CELLS

Korashon L. Watts[a], Vijay Reddy[b†] and Roger Tran-Son-Tay[a,c†]*

[a]Department of Biomedical Engineering, P.O. Box 116131
University of Florida, Gainesville, FL 32611-6131, USA
[b]Division of Hematology/Oncology, P.O. Box 100277,
University of Florida, Gainesville, Florida 32611, USA
[c]Department of Mechanical and Aerospace Engineering, P.O. Box 116250
University of Florida, Gainesville, FL 32611-6250, USA

ABSTRACT

Background: Clinical trials are ongoing involving transplantation of ex vivo expanded hematopoietic stem cells. This study was carried out in order to better understand the effects of ex vivo expansion on the rheology of umbilical cord blood-derived stem cells.

Methods: Micropipette aspiration and recovery experiments were performed using fresh umbilical cord blood CD34+ cells and umbilical cord blood CD34+ cells that were cultured for 1 week. From these experiments, viscosity and cortical tension were determined, and results were analyzed to determine if cultured cells differed from fresh cells in terms of rheology.

Results: As compared to fresh cells, cultured cells had a significantly lower viscosity (221.41±16.88 Pa•s as compared to 515.42±62.87 Pa•s for fresh cells, p=0.0001) and a significantly lower cortical tension (5.54±0.49x10-5 N/m as compared to 2.14±0.23x10-4 N/m for fresh cells, p<0.0001).

---

* Corresponding author: Roger Tran-Son-Tay, Department of Mechanical and Aerospace Engineering, 231 Aerospace Building, P.O. Box 116250. Gainesville, FL 32611-6250. rtst@ufl.edu
† Both authors contributed equally

Conclusions: Cultured and fresh CD34+ cells are likely to behave differently in the body. Cultured cells are likely to flow more easily through the capillaries; however, they may also recover more slowly from a deformation.

Keywords: Cord blood stem cells, ex vivo expansion, rheology, viscosity, cortical tension.

INTRODUCTION

Hematopoietic stem cell transplantation is a treatment option for patients suffering from a wide range of hematopoietic malignancies. Umbilical cord blood (UCB) offers numerous advantages over other sources of transplantable stem cells because of ready availability, less stringent donor-host matching requirements, reduction of risk, and ease of procurement. The first umbilical cord blood transplant occurred in 1990, and since then there have been over 6,000 of these transplants performed worldwide. Over the past decade, significant advances have been made in the field of UCB stem cell transplantation, which have broadened the pool of potential recipients to include adult patients [1] as well as the more traditional pediatric recipients. However, it is known that cell dose (number of transplanted cells per kilogram body weight) continues to be the limiting factor in the success of UCB transplantation [2]. For this reason, clinical trials are ongoing involving transplantation of ex vivo expanded cells as a means of augmenting cell numbers prior to transplant [3, 4]. Advantages offered by using ex vivo expanded cells for transplantation include increased engraftment, reduced risk of infection, an increase in the pool of potential donors, and increased clinical applications [5].

Though preliminary results from clinical trials show promise [6, 7, 8], it has been suggested that culture may affect the functional properties of the cells [5]. For example, it has been shown that ex vivo expanded UCB CD34+ cells differ from fresh UCB CD34+ cells in terms of their differentiation potential [9]. Cultured CD34+ cells were found to generate a significantly higher number of white blood cell progenitors and a significantly lower number of red blood cell progenitors as compared to fresh (uncultured) cells. Therefore, it is important to understand the effects of ex vivo expansion on the properties of CD34+ cells in order to fully consider the health impacts that may result.

One of the governing characteristics of blood cells is the ability to deform, flow through the tiny vessels of the body, and migrate to different regions of the anatomy (for example: sites of infection or tissue damage). This property becomes even more pivotal in ill or immunocompromised patients. Therefore, this study was designed with the goal of determining the effects of ex vivo expansion on the rheological properties of CD34+ cells. To this end, aspiration and recovery experiments were performed on fresh CD34+ cells and cultured CD34+ cells to determine the viscosity and cortical tension of both populations of cells. Aspiration and recovery experiments were performed according to the micropipette technique described by Evans and Yeung [10]. In short, a cell is aspirated under constant pressure into a capillary tube with a diameter less than that of the cell. As the cell flows into the tube, it behaves as a Newtonian liquid and deforms in a time-dependent manner. Following aspiration, the cell is expelled and recovers its spherical shape. By recording various dimensions of the cell during both aspiration and recovery, one can calculate various rheological properties of the cell. This technique has been used for several years to study the properties of various blood cells, include granulocytes [10], neutrophils [11], leukocytes [12], red blood cells [13], platelets [14], and lymphocytes [14, 15].

MATERIALS AND METHODS

Control Population Versus Experimental Population. The "control group" for this study consisted of fresh CD34+ cells, which were pulled from the blood sample immediately after CD34+ purification and prior to aspiration and recovery experiments. The "experimental group" for this study consisted of cultured CD34+ cells, which were expanded for one week and then re-purified for the CD34+ antigen prior to aspiration and recovery experiments. Therefore, cells from both the control and experimental population were handled in the same manner; the only variable between the two populations was the duration of culture (0 days or 7 days).

Cell Sources. Hematopoietic stem cells used in this study were derived from neonatal umbilical cord blood. Samples were obtained from the Stem Cell Lab of Shands Hospital at the University of Florida under an approved IRB protocol with written, informed consent. Umbilical cord blood samples were stored at 4°C and were processed within 36 hours of collection.

Mononuclear Cell Separation. The full volume of umbilical cord blood was first diluted in a 1:1 ratio with 1X phosphate buffered saline (PBS). Mononuclear cells were then separated by centrifugation over the buoyant density solution Lymphoprep (Axis-Shield, Oslo, Norway). The mononuclear cell layer was removed by aspiration and washed with PBS in preparation for CD34+ cell purification.

CD34+ Cell Purification. Following mononuclear cell separation, CD34+ cells were isolated using the MACS magnetic column separation system (Miltenyi Biotec, Auburn, CA). Briefly, CD34+ cells were magnetically labeled using a hapten-conjugated primary monoclonal antibody and an anti-hapten antibody incorporated with MACS magnetic MicroBeads and then run through magnetic columns for positive selection of the CD34+ antigen. Magnetic separation was repeated twice to increase purity.

Flow Cytometry. Cell viability and CD34+ purity were assessed using flow cytometry. Cells were washed and suspended in staining buffer (1X PBS supplemented with 0.5% BSA and 0.01% sodium azide), followed by addition of 10% human AB serum to block non-specific binding. Cells were labeled with a CD34 Phycoerythrin (PE)-conjugated antibody or an isotype matched control, IgG$_1$ Control PE-conjugated antibody (Becton Dickinson, San Diego, CA). As per the instructions of the CD34+ separation system manufacturer (Miltenyi Biotech, Auburn, CA), these antibodies recognized an epitope different from that recognized by the CD34 monoclonal antibody used in the purification step. Cells were then washed twice and stained with Via-Probe fluorescein isothiocyanate (FITC)-conjugated antibody (Becton Dickinson, San Diego, CA) for viability testing. Samples were run on a FACScan flow cytometer and analyzed using CellQuest software to produce plots depicting cell viability and CD34+ cell purity.

Culture of CD34+ Cells. Culture of CD34+ cells was initiated in 24-well plates. Cells were plated at a concentration in the range of 8×10^4 - 4×10^5 CD34+ cells/ml, as determined by preliminary experiments to promote satisfactory growth. Cells were grown in Iscove's

Modified Dulbecco's Medium (IMDM) (GIBCO, Long Island, NY), supplemented with 10% Fetal Bovine Serum (FBS), 5 ng/ml thrombopoietin (TPO), and 50 ng/ml *flt*-3 ligand (FL). All growth factors were obtained from PeproTech, Inc. (Rocky Hill, NJ). Cultures were incubated at 37°C in 100% humidity and 5% CO_2. At intervals of 3-4 days, cultures were re-supplemented with fresh growth factors. Cultures were maintained for 7 days.

Experimental Preparation. Glass capillary tubes (1.0 mm x 0.5 mm, 4") were pulled on a Narishige model PB-7 pipette puller (Narishige Company, Ltd., Tokyo, Japan) to create micropipettes of a desired diameter. Micropipettes used for fresh cell experiments had an average inner diameter of 6.57 μm, and micropipettes used for cultured cell experiments had an average diameter of 7.80 μm. (Cultured cells were found to be 10-20% larger than fresh cells; therefore, micropipette diameters were scaled according.)

Rheology Experiments. An aliquot of "fresh cells" (pulled from the day 0 CD34+ fraction) and an aliquot of "cultured cells" (cultured for 1 week and then re-purified for the CD34+ antigen) were used for aspiration and recovery experiments. Experiments were performed using 10 fresh cells and 10 cultured cells from each of three different umbilical cord blood units, for a total of n=30 fresh cells and n=30 cultured cells. CD34+ cells were suspended in PBS in a clear chamber which was affixed atop of the stage of an Axiovert 100 inverted light microscope. The proximal end of the micropipette was inserted into the cell chamber; the distal end of the micropipette was connected to two coupled water reservoirs in order to provide hydrostatic pressure control. Individual cells were aspirated under constant pressure until the cell was located fully inside the micropipette (Figure 1a). The cell was then held in place for 10 seconds, and expelled from the micropipette. It has been shown that, if the cell is held in place for at least 5 seconds before being expelled, it behaves as a Newtonian liquid with constant cortical tension [11]. However, if the cell is held in place for less than 5 seconds, it undergoes a rapid, elastic rebound behaving as a Maxwell viscoelastic liquid instead [11]. Therefore, a hold period of 10 seconds was selected for these experiments. After expulsion, the cell was observed until it fully recovered its original, undeformed shape (Figure 1b). An attached video system was used to record each experiment, and the Adobe Premiere 6.0 program was used to create still frames at regular time intervals from the recorded experiments. From the still frames, a variety of different dimensions were recorded (Figure 2).

Rheological Analysis. During aspiration experiments, the leading edge of the cell within the micropipette was tracked as a function of time (Figure 1a). From this data, a plot of L_p (the length of the cell inside the micropipette) versus time was generated (Figure 3). Viscosity (μ) values (in units pf Pa•s) were determined by the following equation [16]:

$$\mu = \frac{(\Delta P)D_p}{2(dL_p/dt)m\left(1-\dfrac{1}{\overline{D}}\right)} \tag{1}$$

where ΔP is the pressure (in pascals) as indicated by the height difference between the water reservoirs, D_p is the diameter of the micropipette (in meters), dL_p/dt is the slope of the linear portion of the aspiration curve, $\overline{D}=D_{out}/D_p$, and D_{out} is the diameter of the portion of the cell outside of the pipette. The constant $m \approx 6$, and is derived from the early theory of Evans and Yeung [10].

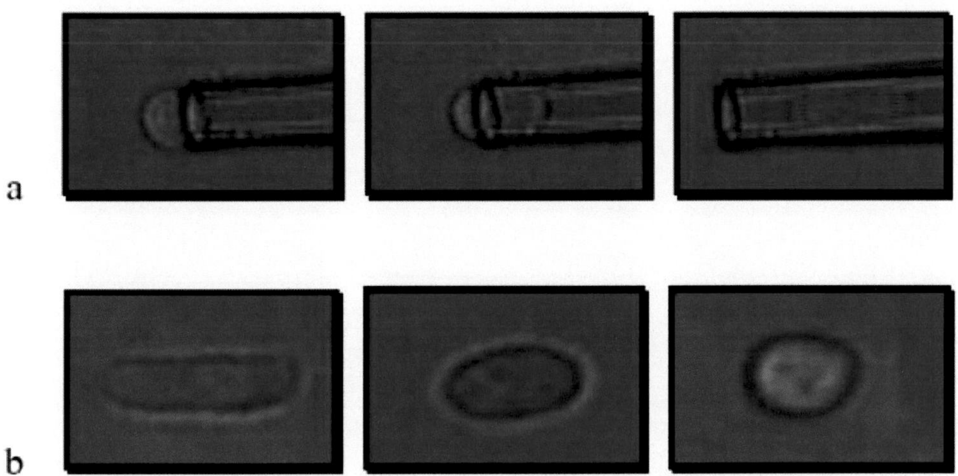

Figure 1. A typical aspiration and recovery experiment, shown at different time points (33X objective and 10X ocular eyepiece). a.) A 10.7-µm CD34+ cell is aspirated into an 8.0-µm pipette. b.) Following expulsion from the micropipette, the CD34+ cell recovers its original, undeformed shape.

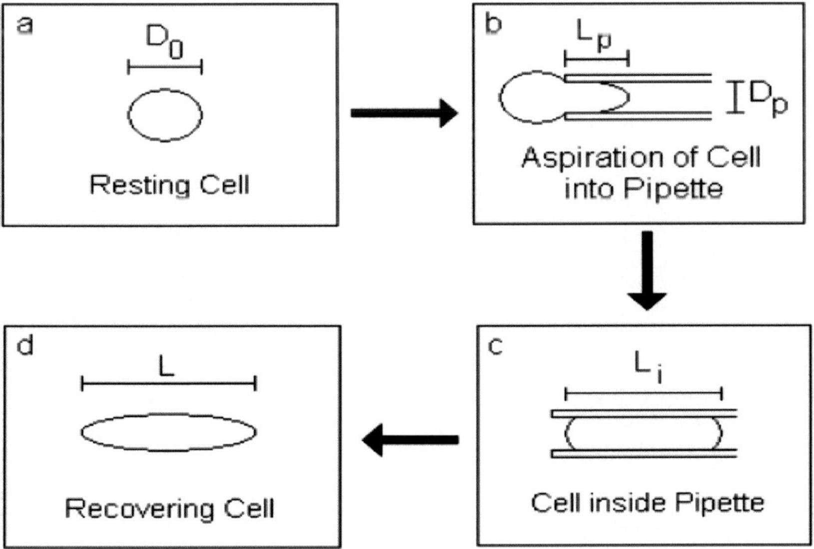

Figure 2. Schematic diagram of aspiration (a, b) and recovery (c, d) experiments, showing the dimensions of interest. D_0 = resting diameter of the cell, L_p = length of the cell inside the micropipette, D_p = diameter of the micropipette, L_i = length of the cell when completely inside the micropipette, L= length of the major axis of the recovering cell.

During recovery experiments, the length of the major axis of the recovering cell was monitored as a function of time (Figure 1b and Figure 4). The equation below governs this relationship for white blood cells [11].

$$\frac{L}{D_0} = \frac{L_i}{D_0} + A\bar{t} + B\left(\bar{t}\right)^2 + C\left(\bar{t}\right)^3 \tag{2}$$

In the above equation, L is the length of the major axis of the recovering cell, D_0 is the diameter of the resting cell, L_i is the length of the cell when completely inside the micropipette, A, B, and C are known function of (L_i/D_0) as defined previously [11] and $\bar{t}$ is a dimensionless time defined by:

$$\bar{t} = \frac{2t(T_0)}{\mu(D_0)} \tag{3}$$

By substituting equation (3) into equation (2) and rearranging the terms, the following cubic equation is obtained:

$$\frac{L_i}{D_0} + A\left(\frac{2tT_0}{\mu D_0}\right) + B\left(\frac{4t^2T_0^2}{\mu^2 D_0^2}\right) + C\left(\frac{8t^3T_0^3}{\mu^3 D_0^3}\right) - \frac{L}{D_0} \tag{4}$$

which can be solved for cortical tension (T_0) in units of N/m.

Statistics. Student's t-tests were used to determine significance between data sets. Differences were taken to be significant at the $p \leq 0.05$ level.

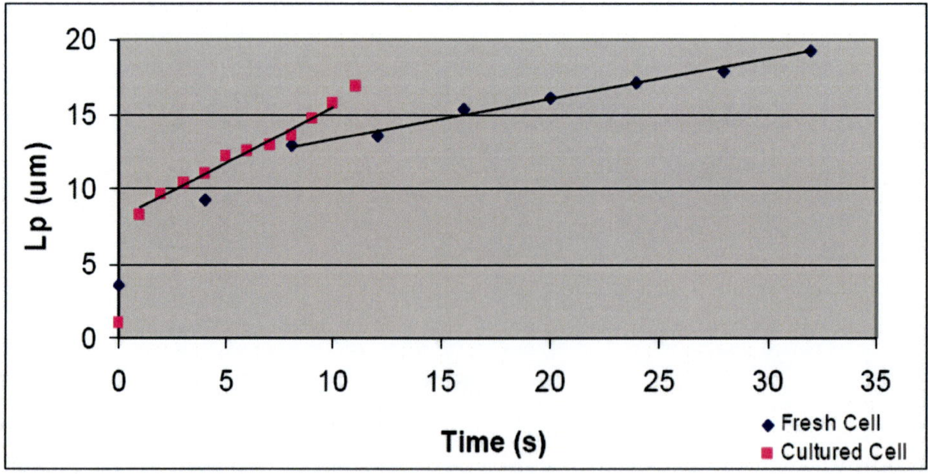

Figure 3. Typical aspiration curves for a fresh CD34+ cell and a cultured CD34+ cell. The cultured CD34+ cell takes much less time to be aspirated into the micropipette, which results in a higher dL_p/dt value. L_p = length of the cell inside the micropipette.

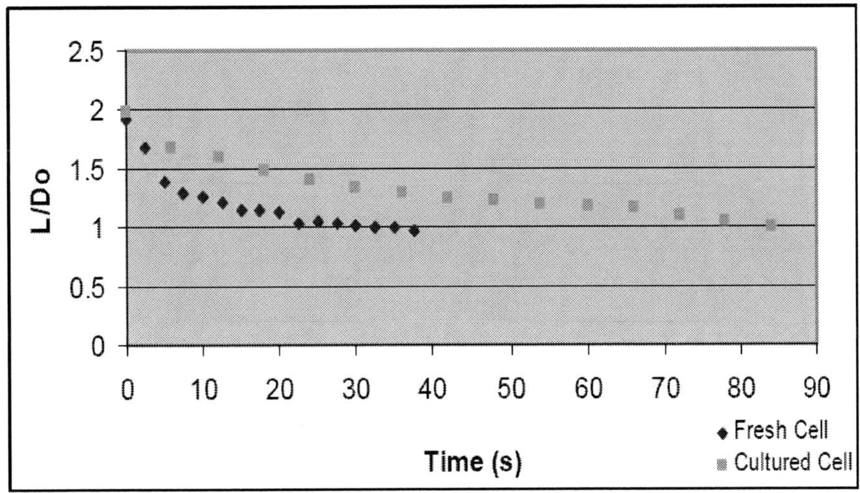

Figure 4. Typical recovery curves for a fresh CD34+ cell and a cultured CD34+ cell. The cultured CD34+ cell takes much longer to recover its original, undeformed shape following expulsion from the micropipette. L=length of the major axis of the recovering cell, D_0=diameter of the resting cell.

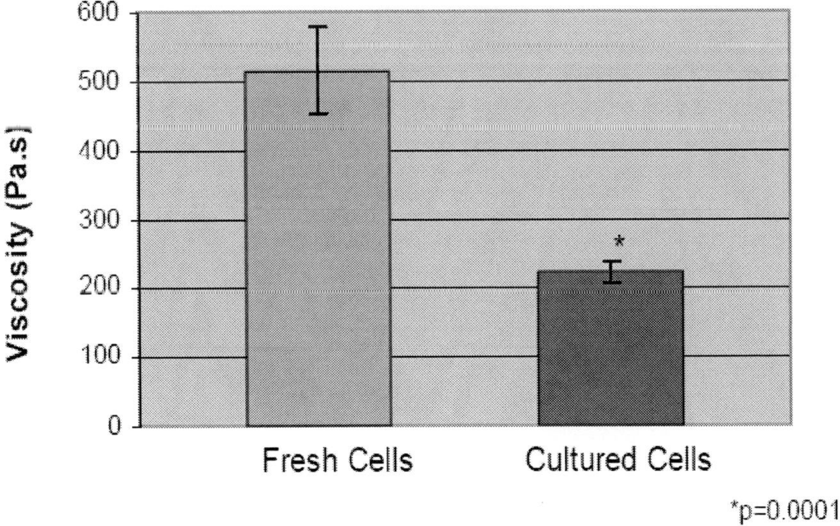

Figure 5. Viscosity of fresh CD34+ cells versus that of cultured CD34+ cells. Fresh cells have an average viscosity of 515.42±62.87 Pa•s, and cultured cells have an average viscosity of 221.41±16.88 Pa•s. Data are shown as mean ± SEM (p=0.0001).

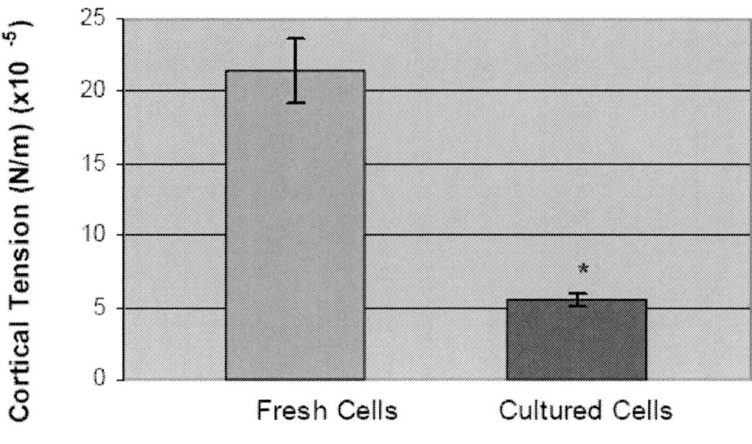

Figure 6. Cortical tension of fresh CD34+ cells versus that of cultured CD34+ cells. Fresh cells have an average cortical tension of $2.14 \pm 0.23 \times 10^{-4}$ N/m, and cultured cells have an average cortical tension of $5.54 \pm 0.49 \times 10^{-5}$ N/m. Data are shown as mean ± SEM (p<0.0001).

Cell Type	μ (Pa·s)	T_0 (N/m)	Reference(s)
CD34+ cells, fresh, human	515.42	2.14×10^{-4}	(present work)
CD34+ cells, cultured, human	221.41	5.54×10^{-5}	(present work)
granulocytes, human	100 - 1000	3.50×10^{-5}	[10]
lymphocytes, human	450.20 - 1081.28	----	[14]
neutrophils, human	5 - 500	2.40×10^{-5} - 3.80×10^{-5}	[10, 17, 18]

Figure 7. Viscosity (μ) and cortical tension (T_0) of various types of white blood cells.

RESULTS

Aspiration and recovery experiments were performed on 30 fresh cells and 30 cultured cells derived from 3 different umbilical cord blood units. As determined by flow cytometry, CD34+ purities obtained from magnetic column separation were in the range of 85-98%. There was no significant difference between the CD34+ percent purity of cells used for "fresh cell" rheology experiments and "cultured cell" rheology experiments.Results showed that the rheological properties of cultured CD34+ cells differ significantly from those of fresh CD34+ cells. Cultured cells were found to have a significantly lower viscosity than cultured cells (p=0.0001). Cultured cells had an average viscosity of 221.41 ± 16.88 Pa·s, while fresh cells had an average viscosity of 515.42 ± 62.87 Pa·s (Figure 5). In addition, cultured cells had a significantly lower cortical tension than fresh cells (p<0.0001). Cultured cells had an average cortical tension of $5.54 \pm 0.49 \times 10^{-5}$ N/m, while fresh cells had an average cortical tension of $2.14 \pm 0.23 \times 10^{-4}$ N/m (Figure 6).

DISCUSSION

This study shows that cultured cord blood-derived CD34+ cells differ from fresh cord blood-derived CD34+ cells in terms of their rheological properties. These differences are manifested by a lower viscosity and a lower cortical tension of cultured cells as compared to fresh cells.

Because lineage-committed cells (i.e. neutrophils, granulocytes, lymphocytes, etc) are phenotypically unique from each other and from the primitive CD34+ progenitor, it in anticipated that their rheological properties will differ as well. In order to exclude lineage-committed cells from aspiration and recovery experiments, cultured cells were re-purified for the CD34+ antigen prior to rheological experiments to ensure that cells used for these experiments were primitive progenitors rather than lineage committed cells. This step served as an internal experimental check point to prevent committed cells from entering the experimental population and interfering with results. The data in Figure 7 lists the viscosity and cortical tension of various types of white blood cells. The variability in the range of values is due to different temperatures, cell treatments, and flow conditions. It is interesting to note that the cortical tension of fresh human CD34+ cells is highest, followed by cultured CD34+ cells, and later by the lineage-committed cells (granulocytes and neutrophils). This is logical, and provides evidence that as the cell proceeds through its life cycle, it loses some of the elasticity of its membrane. In addition, it has been shown that cells that are cultured ex vivo for a period of time tend to alter their levels of antigen expression and gene expression [19-22]. Therefore, it is possible that this phenomenon leads to differences in the physical properties of the cells as well, and these variations may be manifested by a reduced rigidity in the cytoskeleton of cultured cells.

The viscosity of cultured CD34+ cells is around half of the viscosity of fresh CD34+ cells ($\mu_{cultured} \approx 1/2\mu_{fresh}$). This suggests that cultured cells are able to flow more easily through the capillaries of the body than fresh cells. Furthermore, the cortical tension of cultured CD34+ cells is about one fourth of the cortical tension of fresh CD34+ cells ($T_{0,\,cultured} \approx 1/4 T_{0,\,fresh}$). This suggests that cultured cells are less rigid than fresh cells. Consequently, they are likely to recover less quickly from a deformation (see Figure 4). The ratio of cortical tension to viscosity (T_0/μ) is called the recovery time constant [10]. Based on the results above, cultured cells have a recovery time constant that is about half that of fresh cells (as shown in equation 5), explaining why they typically take twice as long to recover from a deformation.

$$\frac{4T_0}{2\mu_{\,cultured}} = \frac{T_0}{\mu_{\,fresh}} \text{ or } \frac{T_0}{\mu_{\,cultured}} = \frac{1}{2}\frac{T_0}{\mu_{\,fresh}} \tag{5}$$

It is important to mention that activated cells are poorly deformable and will not flow into the micropipette [15]. Therefore, it can be stated with certainty that the cells used for the above aspiration experiments were not activated. Furthermore, Miltenyi Biotec states that their magnetic column separation system is cell-friendly, biodegradable, compatible with downstream applications such as cell culture, and that the binding of the antibody to the cell surface does not affect the viability or function of the cell [23-25]. It has also been documented by others that positive selection does not interfere with the properties of CD34+ cells [26, 27]. Therefore, it is unlikely that our results were altered by the isolation process.

These findings have significant implications in the field of hematopoietic stem cell transplantation. CD34+ cells that have been augmented through ex vivo expansion prior to transplantation are likely to behave differently in the body than those that have not been augmented. These results show that ex vivo expanded cells are likely to have less resistance to flow through the capillaries of the body than fresh cells. As blood cells are routinely called upon during the regular function of the hematopoietic and immune systems, this may be beneficial, resulting in more rapid cellular migration and migration-dependent physiological responses.

However, it should also be noted that a high volume of diffusion-governed mass transport occurs during periods when the cell is in direct contact with capillary walls. If the duration of contact is lessened, there is the potential for incomplete exchange between the cell and capillary wall. Furthermore, it is mentioned above that cultured cells are likely to recover more slowly from a deformation. The impact of this finding on patient health is unclear; however, it should be taken into consideration in the post-transplant care of immune-compromised patients.

In conclusion, cultured umbilical cord blood-derived CD34+ cells have different rheological properties than their fresh cell counterparts. These differences will likely cause a variation in the kinetics of cell migration in the body, and therefore should be considered when using ex vivo expanded cells for clinical transplantation.

ACKNOWLEDGMENTS

The authors would like to thank both the Stem Cell Laboratory of Shands Hospital and Cecile Perrault of the Laboratory of Cellular Mechanics and Biorheology at the University of Florida for technical assistance.

REFERENCES

[1]   Tse, W; Laughlin, M. Cord blood transplantation in adult patients. *Cytotherapy,* 2005, 7(3), 228-242.

[2]   Wagner, JE; Barker, JN; DeFor, T, et al. Transplantation of unrelated donor umbilical cord blood in 102 patients with malignant and non-malignant diseases: Influence of CD34 cell dose and HLA disparity on treatment-related mortality and survival. *Blood,* 2002, 100, 1611-1618.

[3]   Shpall, EJ; Quinones, R; Giller, R; et al. Transplantation of ex vivo expanded cord blood. Biol *Blood Marrow Transplant,* 2002, 8(7), 368-376.

[4]   Jaroscak, J; Goltry, K; Smith, A; et al. Augmentation of umbilical cord blood (UCB) transplantation with ex vivo expanded UCB cells: results of a phase 1 trial using the AastromReplicell System. *Blood,* 2003, 101(12), 5061-5070.

[5] Devine, SM; Lazarus, HM; Emerson, SG. Clinical applications of hematopoietic progenitor cell expansion: current status and future prospects. *Bone Marrow Transplant,* 2003, 31, 241-252.

[6] Reiffers,J; Cailliot, C, Dazey, B, et al. Abrogation of post-myeloablative chemotherapy neutropenia by ex vivo expanded autologous CD34-positive cells. *Lancet,* 1999, 354,1092-1093.

[7] McNeice, I; Jones, R; Bearman, SI, et al. Ex vivo expanded peripheral blood progenitor cells provide rapid neutrophil recovery after high-dose chemotherapy in patients with breast cancer. *Blood,* 2000, 96, 3001-3007.

[8] Paquette, RL; Dergham, ST; Karpf, E, et al. Ex vivo expanded unselected peripheral blood: progenitor cells reduce posttransplantation neutropenia, thrombocytopenia, and anemia in patients with breast cancer. *Blood,* 2000, 96, 2385-2390.

[9] Watts, KL; Tran-Son-Tay, R; Reddy, V. The effects of ex vivo expansion on the differentiation of umbilical cord blood-derived CD34+ cells. *Journal of Stem Cells,* 2006, 1(4), xxx-xxx.

[10] Evans, EA and Yeung, A. Apparent viscosity and cortical tension of blood granulocytes determined by micropipette aspiration. *Biophys. J.,* 1989, 56, 151-160.

[11] Tran-Son-Tay, R; Needham, D; Yeung, A; Hochmuth, RM. Time-dependent recovery of passive neutrophils after large deformation. *Biophys. J.,* 1991, 60, 856-866.

[12] Kan, HC; Shyy, W; Udaykumar, HS; Vigneron, P; Tran-Son-Tay, R. Effects of nucleus ot leukocyte recovery. *Annals of Biomedical Engineering,* 1999, 27, 648-655.

[13] Maggakis-Kelemen C; Biselli, M; Artmann, GM. Determination of the elastic shear modulus of cultured human red blood cells. *Biomed. Tech.,* 2002, 47 Suppl 1 Pt 1, 106-109.

[14] Thomas, S; Bolch, W; Kao, KJ; Bova, F; Tran-Son-Tray, R. Effects of x-ray radiation on the rheologic properties of platelets and lymphocytes. *Transfusion,* 2003, 43, 502-508.

[15] Perrault, CM; Bray, EJ; Didier, N; Ozaki, CK, Tran-Son-Tay, R. Altered rheology of lymphocytes in the diabetic mouse. *Diabetologia,* 2004, 47, 1722-1726.

[16] Needham, D and Hochmuth, RM. Rapid flow of passive neutrophils into a 4 μm pipet and measurement of cytoplasmic viscosity. *Journal of Biomechanical Engineering,* 1990, 112, 269-276.

[17] Needham, D and Hochmuth, RM. A sensitive measure of surface stress in the resting neutrophil. *Biophys. J.,* 1992, 61, 1664-1670.

[18] Zhelev, DV; Needham, D; Hochmuth, RM. Role of the membrane cortex in neutrophil deformation in small pipets. *Biophys J,* 1994, 67, 696-705.

[19] Ma, DC; Jin BQ; Sun, YH; Chang, KZ; Dai, B; Chu, JJ; Liu, YG. Changes in cyclin expression during proliferation and differentiation of CD34+ cells derived from fetal liver induced by thrombopoietin. *Sheng Li Xue Bao,* 2001, 53, 296-302.

[20] Denning-Kendall, P; Singha, S; Bradley, B; Hows, J. Cytokine expansion culture of cord blood CD34+ cells induces marked and sustained changes in adhesion receptor and CXCR4 expressions. *Stem Cells,* 2003, 21, 61-70.

[21] Dooley, DC; Oppenlander, BK. Phenotypic and functional analyses of CD34- hematopoietic precursors from mobilized peripheral blood. *Methods Mol. Biol.,* 2004, 263, 201-218.

[22] Li, Q; Cai, H; Liu, Q; Tan, WS. Differential gene expression of human CD34+ hematopoietic stem and progenitor cells before and after culture. *Biotechnol Lett*, 2006, 6, 389-394.

[23] Heath, SL; Tew, JG; Tew, JG; Szakal, AK; Burton GF. Follicular dendritic cells and human immunodeficiency virus infectivity. *Nature*,1995, 26, 740-744.

[24] Turner, J; Dockrell, HM. Stimulation of human peripheral blood mononuclear cells with live Mycobacterium bovis BCG activates cytolytic CD8+ T cells in vitro. *Immunology*, 1996, 87, 339-342.

[25] Parra, E; Wingren, AG; Hedlund, G; Kalland, T; Dohlsten M. The role of B7-1 and LFA-3 in costimulation of CD8+ T cells. *J Immunol*, 1997, 158, 637-642

[26] Lakota, J; Ballova, V; Drgona, L; Durkovic, P; Vranovsky, A. Use of selected CD34+ cells in the treatment of relapsed/progressive HD: experiences from a single center. *Cytotherapy*, 2002, 4, 177-80.

[27] Koutna, I; Klabusay, M; Kohutova, V; Krontorad, P; Svoboda, Z; Kozubek, M; Mayer, J. Evaluation of CD34+ - and Lin- -selected cells from peripheral blood stem cell grafts of patients with lymphoma during differentiation in culture ex vivo using a cDNA microarray technique. *Exp. Hematol*, 2006, 34, 832-840.

In: Stem Cell Research Progress
Editor: Prasad S. Koka, pp. 31-50

ISBN: 978-1-60456-308-5
© 2008 Nova Science Publishers, Inc.

Chapter 3

COMPARISON OF THE DIFFERENTIATION POTENTIAL OF CELL POPULATIONS ISOLATED FROM HUMAN LIPOASPIRATE, TRABECULAR BONE TISSUE AND PERIOSTEAL TISSUE

Klaus Gossens[1], Jan-Thorsten Schantz[1,2],
Dietmar Werner Hutmacher[2,3] and Thiam Chye Lim[1]*
[1] Division of Plastic Surgery, Department of Surgery at NUH of Singapore
[2] Division of Bioengineering National University of Singapore
[3] Department of Orthopedic Surgery, Singapore

ABSTRACT

Recent studies indicate that a variety of cells derived from adult human tissues show the capacity to change their tissue specific differentiation program. In this study we harvested cells from lipoaspirated adipose tissue, trabecular bone and pericranium tissue in order to evaluate and compare the differentiation potential of these populations by exposing them to adipogenic, chondrogenic, osteogenic and neurogenic induction media. Histological and immunochemistry analysis after adipogenic induction revealed adipocyte-like characteristic for all populations. Under osteogenic induction, all populations expressed the bone matrix proteins osteocalcin, osteopontin, osteonectin and secreted alkaline phosphatase, an early bone marker, into the culture medium. The chondrogenic potential was evaluated after staining of acidic glycoconjugates within the extracellular matrix. All induced cell populations were stained positive, indicating the presence of *de novo* cartilage. After neuronal induction, responsive cells exhibited neuron-like morphology with refractile cell bodies, extended long processes terminating in typical growth cones and long axon-like filopodia. Immunocytochemistry revealed the

* **Corresponding Author:** Jan-Thorsten Schantz MD, PhD. Assistant Professor, Division of Plastic Surgery and Division of Bioengineering, National University of Singapore, 5 Lower Kent Ridge Road, National University Hospital, Main Building Level 2. Singapore 119074. Tel: 65 6 772 4226. Fax: 65 6 772 8427. Email: surjts@nus.edu.sg

expression of neurofilament M and neuron specific enolase. Due to the use of heterogeneous cell populations each cell population preferentially expressed the phenotype of the tissue it was isolated from. In comparison; under the respective experimental conditions the periosteal and adipose derived cell populations demonstrated equal versatile developmental potentials according to cell morphology, immunocytochemistry and histology. In contrast the trabecular derived cells displayed a lower differentiation potential. These findings indicate that there are subpopulations within the isolated cell populations allowing all these cell preparations to express adipocyte-, osteoblast-, chondrocyte-, and neuron-like phenotypes.

Keywords: differentiation potential, adipose tissue, trabecular bone, periosteal tissue

INTRODUCTION

To maintain homeostasis, each organ or tissue needs a cell source for replenishment. This source consists of tissue-specific stem cells (TSSC), which are subpopulations of cells capable of self-maintenance and indefinite proliferation potential. TSSC are multipotent, which means, they give rise to a limited number of cell lineages within their normal environment. Recently, studies indicate that TSSC, cultured under appropriate experimental conditions, are able to give rise to cell types normally not present in those tissues in which they were located.

Due to different contents and types of stem or progenitor cells cell populations isolated from different tissues vary in their developmental potential. High developmental potential *in vitro* has been predominantly reported for cells derived from mesenchymal tissues especially from bone marrow (Olmsted-Davis et al., 2003; Schwartz et al., 2002; Gao et al., 2001; Poulsom et al., 2001; Lagaaij et al., 2001; Orlic et al., 2001; Jackson et al., 2001; Krause et al., 2001; Mezey et al., 2000; Alison et al., 2000; Theise et al., 2000; Petersen et al., 1999; Eglitis et al., 1997). In this study we used primary human mesenchymal cell populations derived from human calvarial trabecular bone, human periosteum and human lipoaspirated adipose tissue (PLA). Periosteal and trabecular tissues are thought to contain mesenchymal progenitor cells (MPCs). Cell populations derived from periosteal and trabecular tissues are capable of expressing phenotypes of adipocytes, chondrocytes and osteoblasts (Hutmacher and Sittinger, 2003; Tuli et al., 2003; Ringe et al., 2002; Sottile et al., 2002; Nuttall et al., 1998; Zohar, Sodek, and McCulloch, 1997; Wakitani et al., 1994). Processed PLA-cells could be induced to express adipogenic, chondrogenic, osteogenic, myogenic, and neuro-ectodermal markers (Ugarte et al., 2003; Zuk et al., 2002; Zuk et al., 2001).

Despite extensive characterization of cell populations derived from each of the used compartments, so far there has been no comparative study analyzing the differentiation potential under the same standardized culture conditions. Therefore, the objective of this study was to analyze and compare the differentiation potential of cell populations derived from three different mesenchymal tissue types *in vitro*.

MATERIAL AND METHODS

Harvesting Cell Populations

Human tissue samples were taken according to the guidelines of the Internal Review Board of the National University of Singapore and the National University Hospital. All patients had given written consents for the respective tissue samples to be used for biomedical research. Male patients ageing between 20 to 30 years were included into the study. Trabecular cells were derived from trabecular calvarial bone chips of a patient undergoing routine reconstructive surgery. Periosteal cells were harvested from the pericranium of a patient undergoing reconstructive surgery. For both tissues an explant system was performed (Schantz and Ng; 2004, Gerber and Gwynn, 2001). For this purpose the calvarial bone chips were washed several times with phosphate buffered saline (PBS) and minced manually into small fragments using a bone rongeur. The periosteal tissue was washed in PBS and dissected into the inner cambial layer and the fibrous outer sheet. The cambial layer was minced into smaller pieces. The bone fragments and the pieces of the inner cambial layer were transferred into cell culture flasks, respectively, and covered with culture media. After one week in culture the bone fragments and the pieces of the inner cambial layer were removed and the cultures were washed with media to remove unattached cells. The outgrown plastic adherent cell fraction was used to start the cultures.

Human adipose cells were isolated from ~ 300 mL of lipoaspirated abdominal adipose tissue by washing extensively with PBS. After washing, the tissue was minced mechanically with scissors and digested enzymatically for 45 minutes with 0,075 % (w/v) type I collagenase (Sigma Aldrich; Sigma Aldrich St Louis, USA) in PBS. The digested tissue was centrifuged at 1200 x g for 10 minutes. After resuspending the cell pellet in culture medium the cells were manually counted using trypan blue and seeded into culture flaks. After one day the unattached cells were removed and the plastic adherent cell fraction was used to start the cultures.

Culture Conditions and Expansion

During expansion, cells were maintained in cell culture flasks in a humidified incubator at 37 °C and 5 % CO_2 (Binder, Tuttlingen Germany). Culture medium for adipose and periosteal cells consisted of Dulbecco`s Modified Eagle Medium (DMEM; high glucose), 10 % fetal bovine serum (FBS) and 1 % Penicillin-Streptomycin (P/S). Culture medium for trabecular cells based on M199 Medium supplemented with 10 % FBS and 1 % Penicillin-Streptomycin (P/S) (all from GIBCO; Grand Island, NY). Media were changed twice a week. The cultures were passaged when reaching 70 % confluency to avoid spontaneous differentiation or contact inhibition. All tissue culture plasticware were obtained from Nunc (NalgeNunc, Denmark).

Growth Kinetics

Cells were seeded in 6-well plates at cell densities of 2×10^4 cells cm^2. Daily, three wells were trypsinized and viable cells were counted manually by trypan blue assay using a hemocytometer. A plot of the logarithm of the mean numbers against culture time shows the growth curve of the cell culture. The linear part of the curve represents the logarithmic growth phase. The doubling time was calculated from the linear growth phase with the formula $t_g = \dfrac{t}{n}$ whereas t represents the hours in culture and $n = \dfrac{\log N - \log N_0}{0,3}$ is the generation number (DiGirolamo et al., 1999; Bruder et al., 1997).

Lineage-Specific Induction Conditions and Analysis

To compare the differentiation potentials of cell preparations isolated from the different sources, cells were exposed to adipogenic, osteogenic, chondrogenic and neuronal induction medium (IM). As control, cells were maintained in normal culture medium (NCM). To warrant equal culture conditions, the inductions were performed always at the same time point. For induction studies, cells of the fourth passage were used.

Adipogenic Induction

Cells were seeded on a 24-well plate at densities of 1×10^5 cells per well. After 24 hours the induction was started by replacing the NCM with IM. For adipogenic differentiation, the IM was supplemented with 20 % FBS, 0.5 mM isobutyl-methylxanthine, 1 µM dexamethasone, 10 µM insulin and 200 µM indomethacin (Hauner et al., 1987; Green and Kehinde, 1975). The cells were maintained in IM for 4 weeks. The progress of adipogenic maturation was observed visually by phase-contrast microscopy and histologically by oil red O staining. For staining, cells were fixed with 4 % (w/v) formaldehyde/1 % (w/v) calcium and washed with 70 % ethanol. A counterstaining was performed with Meyers Hematoxyline (Stainings from Sigma Aldrich; Sigma Aldrich St Louis, USA). For quantitation, 10 randomly chosen, non-overlapping low power images of each sample were taken, the positive cells were counted and compared with the total cell number. Additionally, immunostaining of the Peroxisome Proliferators Activated Receptor γ (PPARγ), a transcription factor regulating the peroxisomal fat metabolism, was performed with primary PPARγ antibodies (Chemicon; 1:500). For identification of antigens by light microscopy, we used a commercial immunohistochemistry kit which was used according to the manufactures instructions (DAKO EnVison+System, Peroxidase [DAB] detection system; Dako Cytometrics; Carpinteria, CA).

Osteogenic Induction

For osteogenic induction, the IM was supplemented with 20 % FBS, 0.1 μM dexamethasone, 50 μM ascorbate-2-phosphate and 10 mM β-glycerophosphate (Zuk et al., 2001; Pittenger et al., 1999). The cultures were maintained in IM for 4 weeks. Osteogenic maturation was evaluated by measuring alkaline phosphatase (ALP) and osteocalcin (OC) concentrations in the culture supernatants with quantitative immunoassays (Quidel, Sigma, Biomedical Technologies). We did not assume significant proliferation during osteogenesis and calculated the specific ALP synthesis rate based on the initial cell seeding density of 10^5 cells per well. To determine calcium deposition during osteogenesis, samples were stained with the von Kossa method (Preece, 1972). Immunocytochemistry staining of bone matrix proteins was performed by using anti-OC (osteocalcin), anti-OP (osteopontin) and anti-ON (osteonectin) primary antibodies (all Dako Cytometrics, 1:100).

Chondrogenic Induction

To assess chondrogenic differentiation, cells were maintained in micromass culture (Engstrand, 2003; Zuk et al., 2001; Pittenger et al., 1999; Mackay et al., 1998; Denker et al., 1995). One times 10^6 cells were suspended in 10 μL NCM, seeded in the center of a well of a 48 well plate and were allowed to attach. After 3 hours, cultures were topped up with IM based on DMEM supplemented with 1 % FBS, 1 % P/S, 1 μM insulin, 2.3 nM transforming growth factor-β1 (TGF-β1), 50 nM ascorbate-2-phosphate. The cells were cultured up to two weeks (resulting already in chondrogenic differentiation) with daily media changes. For analysis, the cell aggregates were fixed in 3.7 % (w/v) paraformaldehyde for 15 min at room temperature, washed thoroughly with PBS, cryosectioned and stained for 30 min with 1 % (w/v) Alcian Blue or for 4-6 minutes with 0.1 % (w/v) Safranin O (Sheehan, 1980; Bancroft and Stevens, 1982; Lev and Spicer, 1964).

Neuron-Like Induction

Expression of neuron-like phenotypes were induced by using two similar protocols in simultaneous sister cultures (Safford et al., 2002; Woodbury et al., 2000). For both protocols, CM consisted of DMEM with 10 % FBS, 1 % P/S and 1% Fungizone. According to the Woodbury-Protocol, 24 hours prior to induction the CM was replaced with preinduction medium consisting of DMEM, 20% FBS, 1% P/S, 1% Fungizone and 1mM β-mercaptoethanol. After removal of pre-induction media and washing 3 times with PBS, induction was started by addition of IM (DMEM, 2% DMSO, 200 μM butylated hydroxyanisole [BHA], 1% P/S and 1% Fungizone). After 6 hours, induction media were removed and replaced by maintenance media consisting of DMEM, 200 μM BHA solved in DMSO (2 % final concentration), 25 mM KCl, 2 mM valproic acid, 10 μM forskolin, 1 μM hydrocortisone and 0.8 μM insulin. The Safford protocol implies also a pre-induction media. 24 hours prior to induction, control media were substituted by pre-induction media consisting of DMEM, 20% FBS, 1.1 nM basic Fibroblast Growth Factor (bFGF), 3.2 nM Epidermal

Growth Factor (EGF), 1% P/S and 1% Fungizone. After washing with PBS the induction was started by replacing pre-induction media by induction media. The induction medium was composed of DMEM, 200 μM BHA desolved in ethanol (0,5 % final concentration), 5 mM KCl, 2 μM valproic acid, 10 μM forskolin, 1 μM hydrocortisone, 1% P/S, 1% Fungizone and 0.8 μM insulin. Progress of neuronal maturation was determined by morphological characteristics like the presence of refractile cell bodies and filopodial processes. Samples where fixed in 3,7 % (w/v) paraformaldehyde for 20 minutes and immunostainined with antibodies against neuron specific enolase (NSE; Chemicon 1:5) and neurofilament M (NF-M; Chemicon 1:200).

RESULTS

Cell Doubling Time

The calculated doubling time of the periosteal and trabecular cells was 27 hours ($\pm$ 3 h) respectively, for the adipogenic cells we obtained a doubling time of 33 hours ($\pm$ 4 h). We did not find significant differences in the doubling times for the different cell populations and their cell numbers did not increase significantly during differentiation.

Adipogenic Induction

The first morphological change during adipogenic induction was the formation of intracellular lipid droplets within the cytoplasm, as observed by phase contrast microscopy and further confirmed by oil red O staining (figure 1, left column). Induction of the cell populations with adipogenic IM also caused expression of PPARγ (figure 1, right column). Control cultures grown without adipogenic supplements did not show formation of intracellular lipid droplets or PPARγ expression (figure 1g, h). For the adipose derived cells, the first droplets appeared on day 4 after starting the induction. For the periosteal and trabecular derived cells, the first droplets were observed on day 8 and day 10, respectively. For all populations the average number of intracellular lipid droplets increased with time. The cells displaying lipid droplets exhibited a flat, expanded morphology with the majority of the intracellular volume occupied by lipid vesicles. After 4 weeks in culture, selective staining of the fat vacuoles, positive cells and total cell number were determined. We found that 48 % ($\pm$ 14 %) of the adipose tissue derived cells showed adipocytes-like characteristics. In contrast 10 % ($\pm$ 8 %) of the periosteal cells and 4 % ($\pm$ 2 %) of the trabecular derived cells displayed adipocyte-like morphologies.

Osteogenic Induction

Cells treated with osteogenic supplements produced mineralized matrix which was studied by phase-contrast-microscopy and further confirmed by von Kossa staining, while control cultures did not mineralize (figure 2, right column). The calcified extracellular matrix

(ECM) appeared as black nodules (50-100 μm in diameter) on top of the cell monolayer and could be observed for the trabecular and adipose derived cells on day 4 after induction, in contrast the periosteal derived cells demonstrated first nodule formation on day 6.

Osteocalcin is an osteoblast specific marker (Ducy et al. 2000; Ko et al., 1996; Calvo et al., 1996; Triffitt, 1987). During bone matrix formation OC was secreted into the ECM as well as into the culture medium. Thus, OC concentration in the supernatant can be used as a marker for mineralization. Measurement of the ALP concentration in cultures commonly used as an early marker for bone formation (Ogawa et al., 2004; Jonsson et al., 1999; Tobiume et al., 1997). Hence, in combination with other bone marker proteins, the ALP levels can be used complementarily to determine the progress of bone formation. Figure 3 displays a reciprocal correlation of the ALP and OC concentration in the culture supernatant. For the periosteal and trabecular populations, we observed a biphasic ALP concentration profile with peaking ALP concentrations on day 12. In the last week of culture we measured raising OC and falling ALP concentration within the bone derived cell populations. Periosteal derived cells had a 15 times higher maximal specific ALP synthesis rate (50 pg d^{-1} 10^{-6} cells) than the adipogenic derived (3.4 pg d^{-1} 10^{-6} cells) and a 1.2 times higher rate than the osteogenic derived cells (40 pg d^{-1} 10^{-6} cells). In addition to von Kossa staining the samples were immunostained for the detection of the bone matrix proteins osteocalcin, osteopontin and osteonectin (figure 2, left column; figure 4). Immunocytochemistry and von Kossa staining indicated time-dependent, increasing calcification starting on day 14 for all induced populations.

Whilst the trabecular and periosteal derived populations showed a 2 to 3 fold increase in the ALP levels during osteogenic induction, adipose derived cell populations displayed decreasing ALP levels in the supernatant. Quantitative ELISA and immunocytochemistry also showed a lower OC expression for the adipose derived cells compared to the other populations (figure 3b and figure 2, left column). However, for adipose derived cells immunocytochemistry for ON, OP, von Kossa staining and phase-contrast-microscopy showed initiation of matrix formation on day 14.

Chrondrogenic Induction

High-density micromass cultures incubated in chondrogenic media condensed into small spheroids visible to the naked eye as early as 3 days after initial chondrogenic induction. Over the course of 14 days, the number and size of cartilaginous nodules increased. No nodules were observed in the cultures placed in normal non-chondrogenic culture media. The cells changed their morphology from a spindle shaped fibroblast-like morphology in monolayer cultures to an ellipsoid-spherical morphology in micromass cultures. The nodules formed in the micromass cultures supplemented with chondrogenic media were stained positively for Alcian Blue at pH 2.5 and demonstrated sulfated proteoglycan-rich and well-organized extracellular matrix (ECM) (figure 5, left column).

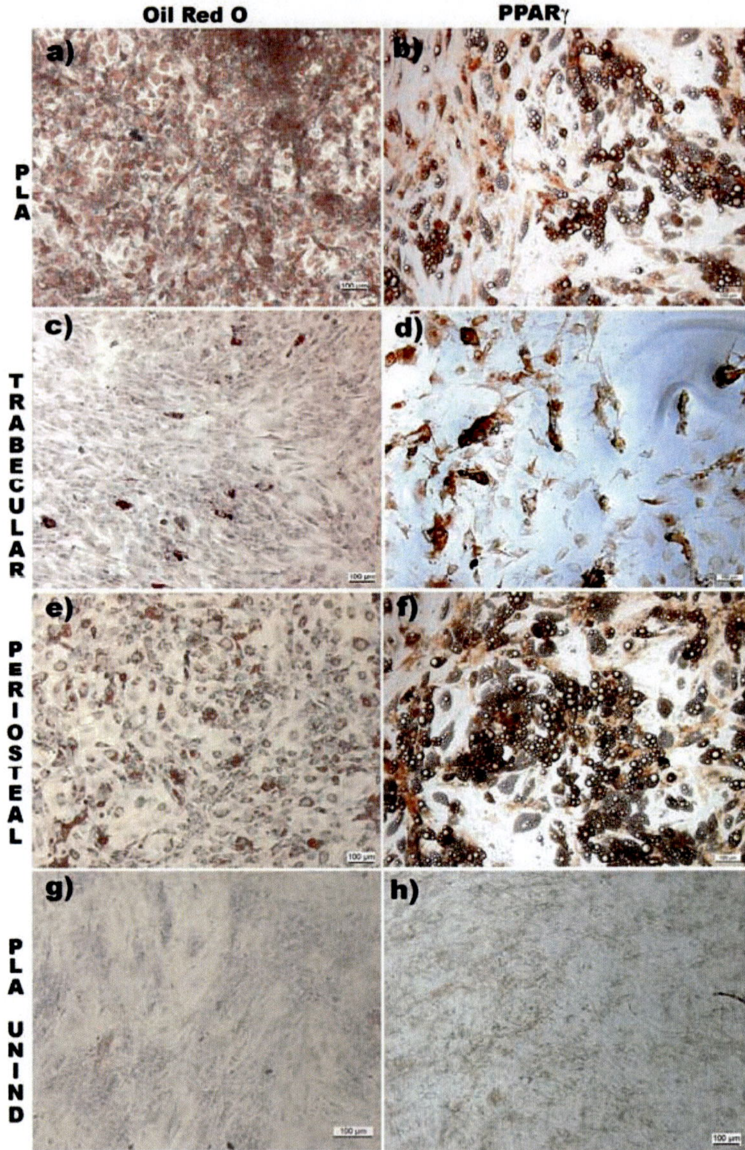

Figure 1. Accumulation of lipid vesicles and expression of PPARγ in adipogenic induced and uninduced cells. All induced cell populations accumulated lipid-filled droplets to a different extent. Undifferentiated processed lipoaspirate (PLA) cells maintained in normal culture medium were used as negative control. After oil Red O staining and counterstaining with Hematoxylin lipid droplets appeared as red spots within a blue stained cytoplasm (left column). After immunostaining of the adipogenic marker PPARγ the cytoplasm of positive cells appeared brown (right column). Uninduced PLA cells showed no significant staining (g, h). UNIND= uninduced.

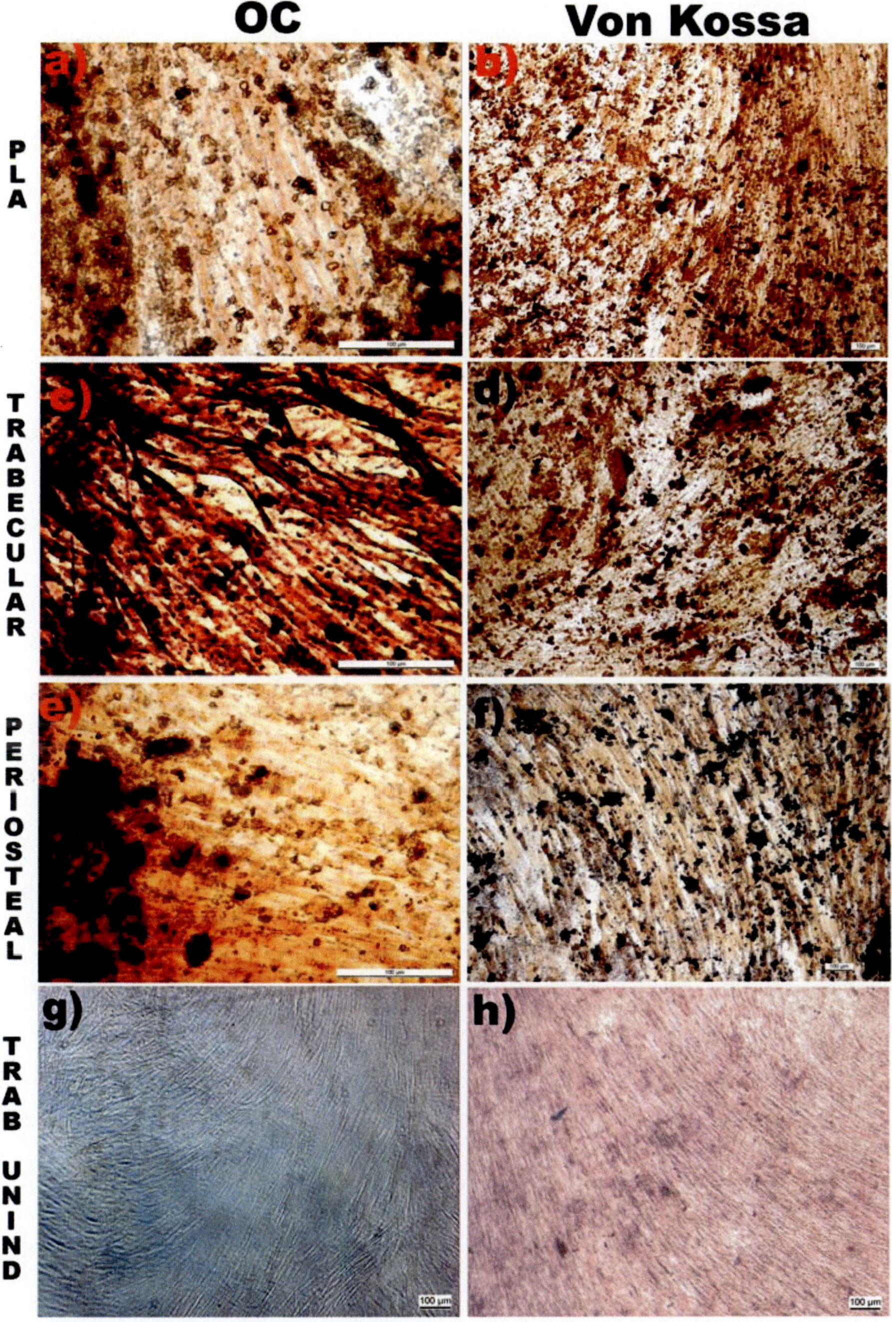

Figure 2. Production of calcified ECM and expression of the bone matrix protein OC in induced and uninduced cell populations. All induced cell populations produced calcified ECM and expressed the matrix protein osteo calcin (OC). Undifferentiated trabecular derived cells maintained in normal culture medium were used as negative control. The presence of a calcified ECM was examined by von Kossa staining (right column). Positive staining could be observed as deep brown nodules. After immunostaining for the extracellular matrix protein OC positive stained regions appeared brown (left column). Negative controls showed no staining (g, h). UNIND= uninduced.

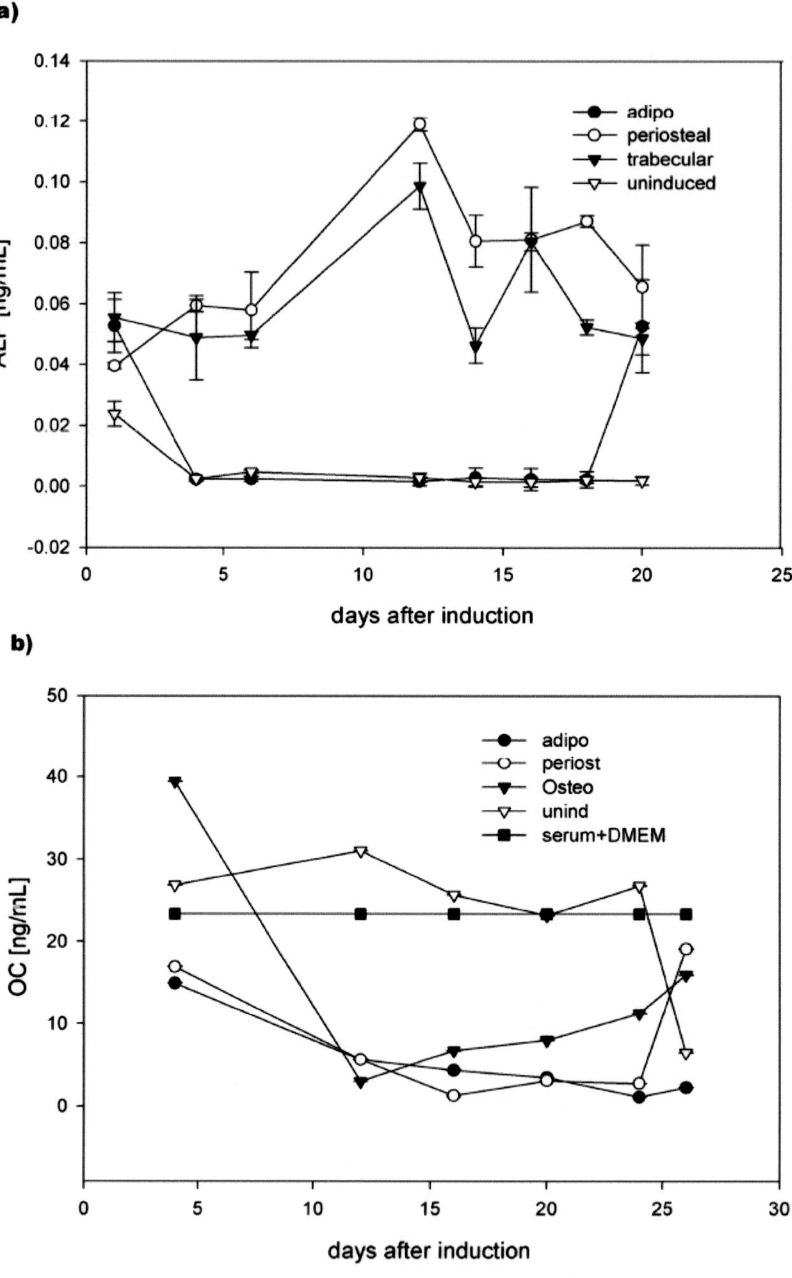

Figure 3. Reciprocal expression of the early bone marker alkaline phosphatase and the late marker osteocalcin during osteogenic inductions. The alkaline phosphatase (a) and osteocalcin (b) concentrations within the culture supernatants were examined by quantitative immunoassays. Due to serum supplementation of the CM, we observed high OC concentrations in the CM and for the uninduced cultures. With continuing osteogenesis the OC was incorporated into the maturing bone matrix and its concentration in the supernatant decreased. During the third week the OC expression rate was higher than the incorporation of the OC and therefore OC accumulated in the culture medium.

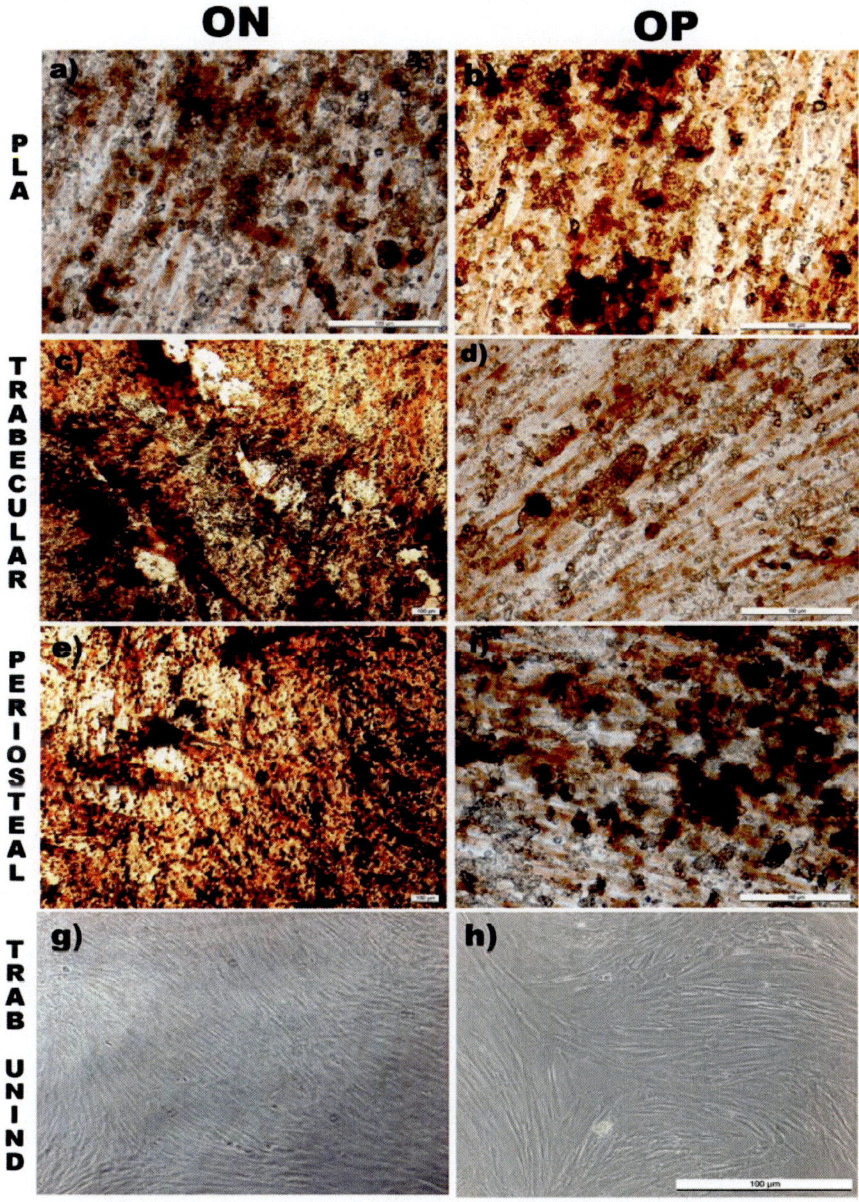

Figure 4. Expression of the bone matrix proteins osteopontin (OP) and osteonectin (ON) in osteogenic induced and uninduced cells. All cell populations produced calcified ECM and expressed the matrix proteins OP and ON after 31 days in osteogenic induction medium. Trabecular derived cells maintained in normal culture medium were used as negative control. The presence of the extracellular matrix proteins ON (left column), OP (right column) was examined after immunostaining. Positive regions appeared as brown nodules. Negative controls showed no staining (g, h). UNIND= uninduced.

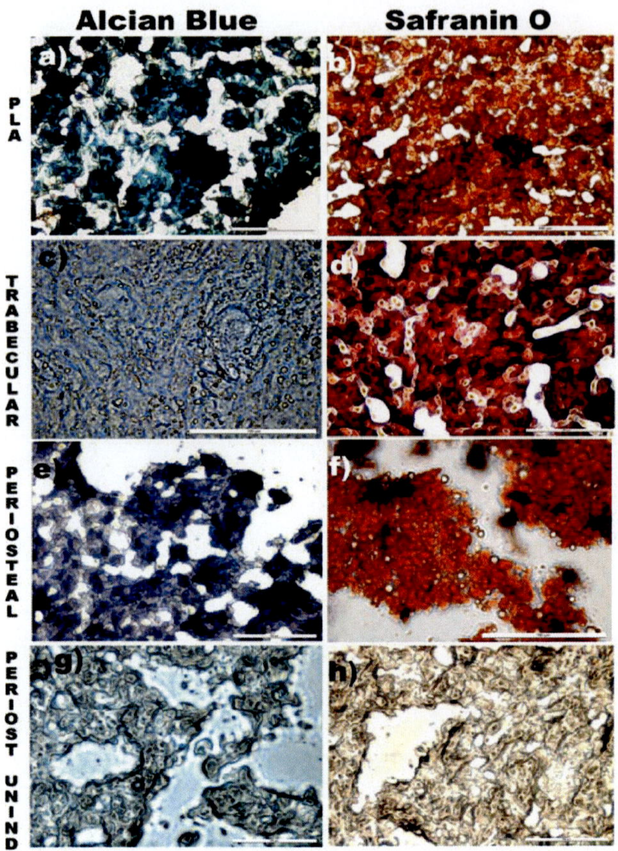

Figure 5. Cell populations cultured in chondrogenic differentiation medium are associated with proteoglycan-rich ECM. The cells were cultured for 2 weeks in differentiation medium using the micromass technique to induce chondrogenesis. Undifferentiated periosteal derived cells were used as negative control. After fixation cells were processed for the presence of sulfated proteoglycans with Alcian Blue (left column) and Safranin O (right column). All populations displayed distinct single cells within the stained copious ECM after 14 days in chondrogenic induction medium (right column; figure 5f, arrows). The negative controls showed no staining. UNIND= uninduced.

Staining of histological sections with Safranin O revealed single morphologically distinct, chondrocyte-like cells embedded in fibrous copious ECM (figure 5, right column). Periosteal cells cultured in CM showed no matrix formation and, therefore, were not stained (figure 5 g, h).

Neuron-Like Induction

Five hours after exposure to the induction medium, responsive cells showed highly significant morphological changes. The cytoplasm retracted towards the nucleus forming a multipolar cell body (figure 6a, arrow). After 24 hours cell bodies became spherical and refractile, showing neuron-like perikaryal characteristics (figure 6b, c).

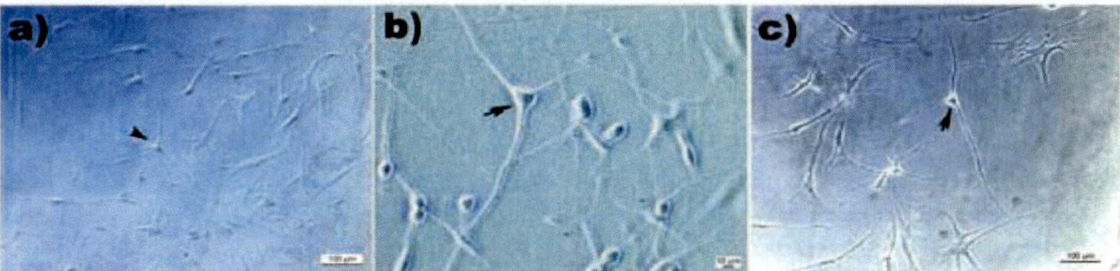

Figure 6. Morphological changes of periosteal derived cells cultured in neuronal induction medium. Cells cultured in neurogenic induction medium for 5 hours retracted the cytoplasm towards the nucleus forming a multipolar cell body (Figure 6a, arrow). After 24 hours cell bodies became spherical and refractile, showing neuron-like perikaryal characteristics (Figure 6b, arrow). The cells started to interconnect and build up networks. After 48 hours neurogenic induction the cells displayed primary and secondary branches, growth cone-like terminal expansions and filopodia-like extensions (Figure 6c, arrow). UNIND= uninduced.

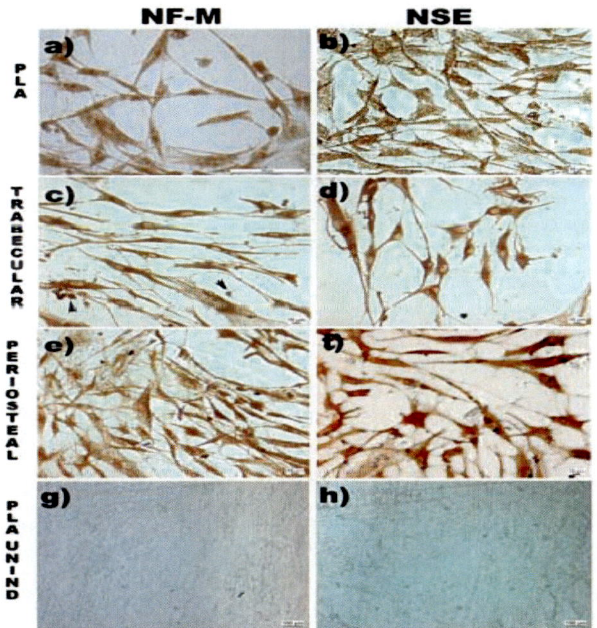

Figure 7. NSE and NF-M expression of neurogenic induced and uninduced cells. Immunostaining of cells cultured in neurogenic medium for 10 hours showed expression of NF-M (left column) and NSE (right column) and elaborated neuron-like morphologies. Undifferentiated cells retained their flattened morphology and did not stain (Figure 7g, h). Differentiated cells stained dark brown for NF-M and NSE expression and displayed multipolar, condensed cell bodies and branched processes. Transitional cells showed retracted, bipolar morphology. The majority of trabecular derived cells showed a transient more bipolar neuronal morphology. Within the first 24 hours of induction all trabecular derived cells rounded up (Figure 7c, arrow) and detached from the surface. Immunocytochemistry with NF-M, a neuron-specific intermediate filament that helps to initiate neurite elongation, stained filamentous structures within the cytoplasm and in the processes brown (left column), whereas NSE stained the entire cytoplasm and the nuclei. With increasing culture time the differentiated cells contacted their surrounding cells via processes building up clusters of neuron-like cells of different morphologies (Figure 7a, b, e). UNIND= uninduced.

After 48 hours the cells developed primary and secondary branches, growth cone-like terminal expansions and filopodia-like extensions (figure 6c). The observed cell body morphologies ranged from simple bipolar cells to large, multipolar cells with highly branched processes. Responsive cells expressed neuron specific enolase and neurofilament M as early as 5 hours after induction, whereas uninduced cells showed neither expression of NSE or NF-M nor morphological changes (figure 7g, h). Neuron-like cells tended to interconnect, building up networks of cells.

Cells within these cell-clusters displayed a strong staining for NSE and survived longer than single, isolated cells (figure 7 f). All trabecular derived cells became rounded, detached and died within the first 24 hours of neurogenic induction (figure 7c, arrows). The periosteal and adipogenic derived populations showed high percentages of dead cells after one day in induction medium. However, the majority (~ 60 %) of the remaining cells possesed neuron-like morphologies and stained positive for NF-M and NSE (figure 7a-f). After 5 days of neurogenic induction also the remaining cells detached and died within the following 48 hours. Due to the high number of dead cells, a detailed quantitative analysis of the neuron-like differentiation resulted in high errors and variances and allowed only semi-quantitative or qualitative descriptions. We also observed that the periosteal cells showed earlier response to induction and maintained an enhanced neuron-like morphology than the adipose derived cells. To assess the potential to differentiate into a neurogenic-phenotype, we used two different protocols according to Woodbury and Safford. In comparison to the Safford protocol the usage of the Woodbury protocol resulted in a higher cell death however in terms of neuron-like morphology better differentiation results.

DISCUSSION

The explant method for bone derived cell populations has been successfully used to obtain new bone formation since the beginning of the last century (Fell, 1934). Previous studies demonstrated that collagenase treatment during the harvesting methodologies reduces the proliferative activity of the cells (Jonsson et al., 1999; Hefley et al., 1981). Hence, we omitted the collagenase step in our explant method. We could not establish an effective explant method for adipose tissue. Therefore, we used a protocol according to Zuk et al. (Zuk et al., 2001).

After exposure to the respective induction media we observed the development of adipocyte-, osteblast-, chondrocyte- and surprisingly neuron-like phenotypes in all induced cell fractions indicating a multipotent differentiation potential for all cell populations.

Adipogenic Induction

Cell populations cultured in adipogenic media showed phenotypic changes in the same manner as shown by a number of other groups (Ogawa et al., 2004; Zuk et al., 2002; Noeth et al., 2001). Percentage values of the positive cells showed high variances, which can be explained with cluster formation of the positive cells. Our results indicate that all three populations are capable of differentiation into adipocyte-like cells to varying degrees.

Osteogenic Induction

All cell populations in osteogenic media exhibited changes in cell structure as reported in the previous studies (Ogawa et al., 2004; Schantz et al., 2003; Zuk et al., 2002; Noeth et al., 2001; Nuttal et al., 1998). After 2 and 4 weeks of osteogenic induction, osteogenesis was detected for all induced populations.

Owen et al (1990) described the developmental sequence of bone in vitro, consisting of three phases: proliferation with matrix secretion, matrix maturation and matrix mineralization. Osteopontin and ALP are early osteogenic markers and characterize the matrix maturation phase. It has been found based on histomorphometrical methods that increased ALP levels correlate with increased bone formation. The expression of OP, an important protein for matrix-cell interaction, was observed 2 weeks after induction. For the bone derived cells the ALP levels were increased around day 12 after induction. Although these results strongly suggest osteogenesis, these proteins are not considered specific markers for osteogenic differentiation. Osteocalcin is a specific bone marker and it determines terminal osteoblast differentiation. Trabecular and periosteal derived cell populations demonstrated increased OC concentrations three weeks after induction. Whereas adipose derived cell populations did not display significant changes in OC and ALP levels in the supernatant during osteogenic induction. Additionally immunocytochemistal analysis of adipose derived preparations confirmed a lower OC expression. Also preliminary RT-PCR data showed no OC mRNA in the osteogenic induced adipose derived cells (unpublished data). It was previously reported that dexamethasone inhibits the expression of osteocalcin in processed lipoaspirate (PLA) cells (Zuk et al., 2002). Since we used dexamethasone in our IM, the low OC concentration in the supernatant of the adipose derived populations could be explained by this phenomenon.

Chondrogenic Induction

Micromass culture technique combined with media supplementation with TGF-β1 resulted in the condensation of the cells into three-dimensional aggregates which has also been shown by studies from several other groups (Zuk et al., 2001; Noeth et al., 2001; Pittenger et al., 1999; Mackay et al., 1998; Denker et al., 1995). These aggregates stained positively for Alcian Blue and Safranin O. Alcian Blue and Safranin O staining is specific for acidic, highly sulfated polysaccharides, glycosaminglycans and glycoproteins within the ECM. Since cartilage has a high content of acidic glycosaminglycans and acidic chondroitin sulfates enhanced staining of cell populations after induction indicate cartilage formation for all induced populations.

Neuron-Like Induction

In this study we used two *in vitro* differentiation protocols for sister cell preparations. Despite high numbers of dead cells we surprisingly observed neuron-like morphologies in all induced populations. However, the cells did not maintain their neuron-like phenotype for more than 7 days and trabecular-derived cells died within the first 24 hours of induction. Like studies performed by Woodbury et al. (2000) we also detected expression of the neuron specific proteins NSE and NF-M within 5 hours after the commencement of induction. NSE staining was enhanced in cell clusters indicating a positive developmental impact of the cells among each other. Our results demonstrate that human adipose, periosteal and trabecular derived cell populations cultured in the presence of specific soluble mediators possess the ability to express neuronal markers and show morphologic changes analogous to neuronal phenotypes.

Since we used heterogeneous cell cultures, the underlying mechanisms, which are responsible for the observed differentiation potential, must comprise different proportions of stem and precursor cells that are capable of differentiating into mesenchymal and non-mesenchymal lineages. These cultures most likely contain partially committed precursors with various degrees of maturation, and mature non-proliferative cells that dedifferentiate when placed in culture (Jonsson et al., 1999). Another possibility could be that the cells of each induced lineage were already present in the initial cell preparation and that changing culture conditions due to the induction could positively select these "contaminating cells" and support their proliferation and induce the expression of lineage specific proteins. For instance Zuk et al found in primary cultures of adipose tissue besides MSCs, small numbers of hematopoietic cells, pericytes, endothelial cells and smooth muscle cells. But they state that the frequency of these contaminating cells which were predominantly mature non-proliferative cells appears to diminish quickly through serial passages in culture and, therefore, do not influence the differentiation of the proliferating multipotent cells (Zuk et al., 2002). Early lineage specific staining (2-4 days after induction) of our induced cultures showed no positive results, supporting the argument that there are hardly any contaminating mature cells within the preparations left after serial passaging in normal non-inductive culture media.

Due to the use of heterogeneous populations, each cell population differentiated preferentially into the lineage it was isolated from. Under the applied conditions the periosteal and adipose derived cell populations showed equally versatile differentiation potentials according to cell morphology, immunocytochemistry and histology. Whereas the trabecular derived cells displayed a lower differentiation potential. Adipose and periosteal tissues are highly vascularized tissues possibly explaining their higher differentiation potential due to increased numbers of progenitor cells compared to the less vascularized trabecular tissue.

Our observations suggest that all three sources have the capacity to express mesenchymal and non-mesenchymal derivatives, showing that intrinsic genomic mechanisms of commitment, lineage restriction and cell fate might be mutable. However, it is important to note that the cells displayed their specific differentiation potential only in vitro, a phenomenon which they might not show in vivo. Furthermore one must consider epigenetic changes of isolated cells resulting in extended differentiation behaviour.

ACKNOWLEDGMENTS

This study was supported by grants from the National Health Care Group Singapore and the National Medical Research Council Singapore (NMRC) grant No. NHG-RPR/02081 and grant No. NMRC/0747/2003. The authors thank Prof. H-G. Machens from the Department of

Plastic and Reconstructive Surgery, University of Luebeck, Germany, for his help to review the manuscript.

REFERENCES

[1]   Alison MR, Poulsom R, Jeffery R, Dhillon AP, Quaglia A, Jacob J, Novelli M, Prentice G, Williamson J, Wright NA. (2000) Hepatocytes from non-hepatic adult stem cells. *Nature* 406(6793):257.

[2]   Bancroft J. and A. Stevens (1982) *Theory and Practice of Histological Technique,s* Batelle Press Ohio 2nd edition, pp 163, 173-174.

[3]   Bruder SP, Jaiswal N, Haynesworth SE. (1997) Growth kinetics, self-renewal, and the osteogenic potential of purified human mesenchymal stem cells during extensive subcultivation and following cryopreservation. *J. Cell Biochem..* 64(2):278-94.

[4]   Calvo MS, Eyre DR, Gundberg CM. (1996) Molecular Basic and Clinical Application of biological Markers of Bone Turnover. *Endocr. Rev.* 17(4):333-68.

[5]   Denker AE, Nicoll SB, Tuan RS. (1995) Formation of cartilage-like spheroid by micromass culture of murine C3H10T1/2 cells upon treatment with transforming growth factor beta-1. *Differentiation.* 59(1):25-34.

[6]   DiGirolamo CM, Stokes D, Colter D, Phinney DG, Class R, Prockop DJ. (1999) Propagation and senescence of human marrow stromal cells in culture: a simple colony-forming assay identifies samples with the greatest potential to propagate and differentiate. *Br. J. Haematol.* 107(2):275-81.

[7]   Ducy P, Schinke Th, Karsenty G. (2000) The osteoblast: a sophisticated fibroblast under central surveillance. *Science.* 289.

[8]   Eglitis MA, Mezey E. (1997) Hemopoetic cells differentiate into both microglia and macroglia in the brains of adult mice. *Proc. Natl. Acad. Sci. USA.* 94(8):4080-5.

[9]   Engstrand T. (2003) Molecular biologic aspects of cartilage and bone: potential clinical applications. *Ups. J. Med. Sci.* 108(1):25-35.

[10]   Fell HB, Robison R (1934) The development of the calcifying mechanism in avian cartilage and osteoid tissue. *Biochem. J.* 28, 2243-2253.

[11]   Gao Z, McAlister VC, Williams GM. (2001) Repopulation of liver endothelium by bone marrow-derived cells. Lancet 357(9260):932-3.

[12]   Gerber I., Gwynn I. (2001). Influence of cell isolation, cell culture density, and cell nutrition on differentiation of rat calvarial osteoblast-like cells in vitro. Microenvironment of osteoblast-like cells *in vitro European Cells and Materials* Vol. 2. 2001 (pages 10-20) ISSN 1473-2262.

[13]   Green H and Kehinde O. (1975) An established preadipose cell line and its differentiation in culture. II. Factors affecting the adipose differentiation. *Cell.* 5(1):19-27.

[14]   Hauner H, Schmid P, Pfeiffer EF. (1987) Glucocortiods and insulin promote the differentiation of human adipocyte precursor cells into fat cells. *J. Clin. Endocrinol. Metab.* 64(4):832-5.

[15]   Hefley T, Cushing J, Brand J (1981) Enzymatic isolation of cells from bone: cytotoxic enzymes of bacterial collagenase. *Am. J. Physiol.* 240, 234-238.

[16] Hutmacher DW. and Sittinger M. (2003) Periosteal Cells in Bone Tissue Engineering. *Tissue Engineering* 9 (4) Suppl.1: 45-64.

[17] Jackson KA, Majka SM, Wang H, Pocius J, Hartley CJ, Majesky MW, Entman ML, Michael LH, Hirschi KK, Goodell MA. (2001) Regeneration of ischemic cardiac muscle and vascular endothelium by adult stem cells. *J. Clin. Invest.* 107(11):1395-402.

[18] Jonsson K.B., Frost A., Nilsson O., Ljnunghall S., Ljunggren Ö. (1999). Three isolation techniques for primary culture of osteoblast-like cells. *Acta Orthop. Scand.* 70(4):365-373.

[19] Ko SC, Cheon J, Kao C, Gotoh A, Shirakawa T, Sikes RA, Karsenty G, Chung LW (1996) Osteocalcin promoter-based toxic gene therapy for the treatment of osteosarcoma in experimental models. *Cancer Res.* 56(20):4614-9.

[20] Krause DS, Theise ND, Collector MI, Henegariu O, Hwang S, Gardner R, Neutzel S, Sharkis SJ. (2001) Multiorgan, multi-lineage engraftment by a single bone marrow-derived stem cell. *Cell.* 105(3):369-77.

[21] Lagaaij EL, Cramer-Knijnenburg GF, van Kemenade FJ, van Es LA, Bruijn JA, van Krieken JH. (2001) Endothelial cell chimerism after renal transplantation. *Lancet.* 357(9249):33-7.

[22] Lev R and Spicer SS. (1964) Specific staining of sulfate groups with Alcian Blue at low pH. *J. Histochem. Cytochem.* 12:309.

[23] Mackay AM, Beck SC, Murphy JM, Barry FP, Chichester CO, Pittenger MF. (1998) Chondrogenic differentiation of cultured human mesenchymal stem cell from marrow. *Tissue Eng.* 4(4):415-28.

[24] Mezey E, Chandross KJ, Harta G, Maki RA, McKercher SR. (2000) Turning blood into brain: cells bearing neuronal antigens generated in vivo from bone marrow. *Science.* 290(5497):1779-82.

[25] Noeth U, Osyczka AM, Tuli R, Hickok NJ, Danielson KG, Tuan RS. (2001) Multilineage mesenchymal differentiation potential of human trabecular bone-derived cells. *J. Orthop. Res.* 20(5):1060-9.

[26] Nuttall ME, Patton AJ, Olivera DL, Nadeau DP, Gowen M. (1998) Human Trabecular bone Cells are able to Express both Osteoblastic and Adipocytic Phenotype: Implications for Osteopenic Disorders. *J. Bone Miner Res.* 13(3):371-82.

[27] Ogawa R, Mizuno H, Watanabe A, Migita M, Shimada T, Hyakusoku H. (2004) Osteogenic and chondrogenic differentiation by adipose-derived stem cells harvested from GFP transgenic mice. *Biochem. Biophys. Res. Commun.* 313(4):871-7.

[28] Olmsted-Davis Elizabeth A. , Zbigniew Gugala , Fernando Camargo,Francis H. Gannon, KathyJo Jackson, Kirsten Anderson Kienstra , H. David Shine, Ronald W. Lindsey , Karen K. Hirschi, Margaret A. Goodell, Malcolm K. Brenner, Alan R. Davis (2003) Primitive adult hematopoietic stem cells can function as osteoblast precursors. *PNAS*, 100: 15877-15882.

[29] Orlic D, Kajstura J, Chimenti S, Jakoniuk I, Anderson SM, Li B, Pickel J, McKay R, Nadal-Ginard B, Bodine IM, Leri A, Anversa P. (2001) Bone marrow cells regenerate infracted myocardium. *Nature.* 410(6829):701-5.

[30] Owen TA, Aronow M, Shalhoub V, Barone LM, Wilming L, Tassinari MS, Kennedy MB, Pockwinse S, Lian JB, Stein GS. (1990) Progressive development of the rat osteoblast phenotype in-vitro: reciprocal relationships in the expression of genes

associated with osteoblast proliferation and differentiation during formation of the bone extracellular matrix. *J. Cell Physiol.* 143(3):420-30.

[31] Petersen BE, Bowen WC, Patrene KD, Mars WM, Sullivan AK, Murase N, Boggs SS, Greenberger JS, Goff JP. (1999) Bone marrow as a potential source of hepatic oval cells. *Science.* 284(5417):1168-70.

[32] Pittenger MF, Mackay AM, Beck SC, Jaiswal RK, Douglas R, Mosca JD, Moorman MA, Simonetti DW, Craig S, Marshak DR. (1999) Multilineage potential of adult human mesenchymal stem cells. *Science.* 284(5411):143-7.

[33] Poulsom R, Forbes SJ, Hodivala-Dilke K, Ryan E, Wyles S, Navaratnarasah S, Jeffery R, Hunt T, Alison M, Cook T, Pusey C, Wright NA. (2001) Bone marrow contributes to renal parenchymal turnover and regeneration. *J. Pathol.* 195(2):229-35.

[34] Preece A. (1972) *A Manual for Histologic Technicians*, 3rd edition. Little, Brown, Boston.

[35] Ringe J, Kaps C, Burmester GR, Sittinger M. (2002) Stem cells for regenerative medicine: advances in the engineering of tissues and organs. *Naturwissenschaften.* 89(8):338-51.

[36] Safford KM, Hicok KC, Safford SD, Halvorsen YD, Wilkison WO, Gimble JM, Rice HE. (2002) Neurogenic differentiation of murine and human adipose-derived stromal cells. *Biochem. Biophys. Res. Commun.* 294(2):371-9.

[37] Sheehan D.(1987) *Theory and Practice of Histotechnology.* Batelle Press, p. 226-227.

[38] Schantz JT, Teoh SH, Lim TC, Endres M, Lam CX, Hutmacher DW. (2003) Repair of calvarial defects with customized tissue-engineered bone grafts I. Evaluation of osteogenesis in a three dimensional culture system. *Tissue Eng.;* 9 Suppl 1: S113-26.

[39] Schantz JT. and NG KW. (2004) *A Manual for Primary Human Cell Culture*, World Scientific Publisher, NJ.

[40] Schwartz RE, Reyes M, Koodie L, Jiang Y, Blackstad M, Lund T, Lenvik T, Johnson S, Hu WS, Verfaillie CM. (2002) Multipotent adult progenitor cells from bone marrow differentiate into functional hepatocyte-like cells. J. Clin. Invest. 109(10):1291-302.

[41] Sottile V, Halleux C, Bassilana F, Keller H, Seuwen K. (2002) Stem cell characteristics of human trabecular bone-derived cells. *Bone* 30(5):699-704.

[42] Theise ND, Nimmakayalu M, Gardner R, Illei PB, Morgan G, Teperman L, Henegariu O, Krause DS. (2000) Liver from bone marrow in humans. *Hepatology.* 32(1):11-6.

[43] Tobiume H, Kanzaki S, Hida S, Ono T, Moriwake T, Yamauchi S, Tanaka H, Seino Y. (1997) Serum Bone Alkaline Phosphatase Isoenzyme Levels in Normal Children and Children with Growth Hormone (GH) Deficiency: A Potential Marker for Bone Formation and Response to GH Therapy. *J. Clin. Endocrinol. Metab.* 82(7):2056-61.

[44] Triffitt JT. (1987) The Special Proteins of Bone Tissue. *Clin. Sci* .(Lond). 72(4):399-408.

[45] Tuli R, Tuli S, Nandi S, Wang ML, Alexander PG, Haleem-Smith H, Hozack WJ, Manner PA, Danielson KG, Tuan RS. (2003) Characterization of mesenchymal progenitor cells derived from human trabecular bone. *Stem Cells.* 21(6):681-93.

[46] Ugarte DA, Morizono K, Elbarbary A, Alfonso Z, Zuk PA, Zhu M, Dragoo JL, Ashjian P, Thomas B, Benhaim P, Chen I, Fraser J, Hedrick MH. (2003) Comparison of Multi-Lineage Cells from Human Adipose Tissue and Bone Marrow. *Cells Tissues Organs.* 174(3):101-9.

[47] Wakitani S, Goto T, Pineda SJ, Young RG, Mansour JM, Caplan AI, Goldberg VM. (1994) Mesenchymal cell-based repair of large, full-thickness defects of articular cartilage. *J. Bone Joint Surg. Am.* 76(4):579-92.

[48] Woodbury D, Schwarz EJ, Prockop DJ, Black IB. (2000) Adult rat and human marrow stromal cells differentiate into neurons. *J. Neurosci. Res.* 61(4):364-70.

[49] Zohar, Sodek and McCulloch (1997) Characterization of stromal Progenitor Cells Enriched by Flow Cytometry. *Blood* 90:3471-3481.

[50] Zuk PA, Zhu M, Mizuno H, Huang J, Futrell JW, Katz AJ, Benhaim P, Lorenz HP, Hedrick MH. (2001) Multinlineage cells from human adipose tissue: implications for cell-based therapies. *Tissue Eng.* 7(2):211-28.

[51] Zuk PA, Zhu M, Ashjian P, De Ugarte DA, Huang JI, Mizuno H, Alfonso ZC, Fraser JK, Benhaim P, Hedrick MH. (2002) Human adipose tissue is a source of multipotent stem cells. *Mol. Biol. Cell.* 13(12):4279-95.

In: Stem Cell Research Progress
Editor: Prasad S. Koka, pp. 51-58

ISBN: 978-1-60456-308-5
© 2008 Nova Science Publishers, Inc.

Chapter 4

ADULT NEURAL STEM CELLS AND THEIR NICHES: DEVELOPMENTAL AND THERAPEUTIC IMPLICATIONS

*Philippe Taupin**

National Neuroscience Institute, Singapore
National University of Singapore
Nanyang Technological University, Singapore

ABSTRACT

Contrary to long-held dogma, neurogenesis occurs in the adult brain and neural stem cells (NSCs) reside in the adult central nervous system (CNS). This suggests that the CNS has the potential to self-repair. In support of this contention, neurogenesis is modulated in the adult brain, new neuronal cells are generated at sites of degeneration and injuries, after transplantation adult-derived neural progenitor and stem cells adopt the fate of the tissue, and adult stem cells may have the potential to differentiate into other lineages. Niches are microenvironments that control stem cell activity. The therapeutic potential of stem cells, particularly adult NSCs, lies therefore in the interaction between stem cells and their niches. Neurogenic niches have been identified in the adult brain. Unraveling the interactions between NSCs and their niches will contribute to our understanding of the developmental potential of adult stem cells, and to bring adult NSC research to therapy.

Keywords: neurogenesis, multipotent, pluripotent stem cells, transdifferentiation.

INTRODUCTION

Neurogenesis occurs primarily in two areas of the adult brain, the subventricular zone (SVZ) and dentate gyrus (DG) of the hippocampus, in various species including human [1-3].

---

* **Correspondence:** 11 Jalan Tan Tock Seng, Singapore 308433. Tel. (65) 6357 - 7533. Fax (65) 6256 - 9178. Email: obgpjt@nus.edu.sg

In the DG, newly generated neuronal cells in the subgranular zone (SGZ) migrate to the granular layer, where they differentiate into granule-like cells [4]. In the SVZ, cells are generated in the anterior part of the SVZ and migrate, trough the rostro-migratory stream (RMS), to the olfactory bulb (OB) where they differentiate into interneurons [5, 6]. In the spinal cord gliogenesis, but not neurogenesis, occurs throughout adulthood [7]. It is hypothesized that newly generated neuronal cells originate from stem cells in the adult brain [1, 2]. Astrocytes expressing the neurofilament nestin are reported to be at the origin of newly generated neuronal cells in the SVZ and DG, suggesting that a population of glial cells may correspond to stem cells of the adult brain [8-10].

In vitro, self-renewing multipotent NSCs have been isolated and characterized from various regions of the adult CNS, including the spinal cord, supporting the existence of NSCs in the CNS [2].

Neurogenesis is modulated in the DG and OB by a broad range of stimuli, like environmental enrichment, tropic factors and cytokines, neurotransmitters and various physiopathological conditions [11, 12]. Newly generated neuronal cells are generated at sites of degeneration in the diseased and injured brain, as in Huntington's disease and after experimental strokes [13]. In this latter case, cell tracking studies reveal that neural progenitor and stem cells originate from the SVZ; they migrate partially through the RMS to the sites of degeneration [14, 15]. This suggests the involvement of adult neurogenesis in a broad range of functions, yet to be fully understood, and in a regenerative attempt [13]. The confirmation that neurogenesis occurs in the adult brain and NSCs reside in the adult CNS suggests that it has the potential for self-repair. Cellular therapy in the CNS would involve the stimulation or transplantation of neural progenitor and stem cells.

In adult tissues, stem cell activity, self-renewal and multipotentiality, is regulated by microenvironments in specialized structures, yet to be defined, called niches [16]. Stem cell niches have been identified in the skin, bone marrow, liver, and gut [17-21]. Recent evidences show that niches for neurogenesis exist, and control NSC activity in the adult brain [22]. In this manuscript, we will review the evidences that neurogenic niches exist in the adult brain, their identities, and role in the developmental potential of adult NSCs. Finally, we will discuss their implication for stem cell therapy.

NEUROGENIC NICHES EXIST IN THE ADULT CNS

In the adult hippocampus, newly generated neuronal cells differentiate into neuronal, and glial cells [4]. In the adult spinal cord, newly generated neuronal cells differentiate exclusively into glial lineages [7]. Self-renewing multipotent NSCs have been isolated and characterized in vitro from various region of the adult CNS, including the hippocampus, spinal cord and retina [23-25]. Upon transplantation in rodents, adult rat hippocampal-derived neural progenitor and stem cells survive, migrate and differentiate into neuronal and glial lineages in the adult hippocampus, whereas rat spinal cord-derived neural progenitor and stem cells differentiate into glial lineages only in the cord, and human retinal stem cells differentiate into neural retinal cells, particularly photoreceptors, in the eye [26-29]. This show that neural progenitor and stem cells isolated from the adult CNS behave like endogenous progenitor cells from the region from where they are derived.

In heterotypic transplantation, rat spinal cord-derived neural progenitor and stem cells differentiate into neuronal and glial lineages in the adult hippocampus, and adult rat hippocampal-derived neural progenitor and stem cells adopt morphologies similar to those of neuronal and astroglial cells of the retina, but do not express mature retinal markers [27, 30]. This shows that the microenvironment of the spinal cord and hippocampus controls the fate of spinal cord-derived neural progenitor and stem cells, and that adult rat hippocampal-derived neural progenitor and stem cells elicit limited capacity to differentiate into mature retinal phenotypes. The evidence that spinal cord-derived neural progenitor and stem cells differentiate into neuronal and glial lineages when grafted into the adult hippocampus, and into glial lineages only when grafted into the spinal cord, suggests the existence of neurogenic niches in the adult CNS.

NEUROGENIC NICHES OF THE ADULT BRAIN

Two niches for neurogenesis have been identified in the adult brain: an angiogenic and astroglial niches. In the SGZ, newly generated neuronal cells are associated with the formation of new blood vessels, and endothelial cells release soluble factors that maintain NSC self-renewal [31, 32]. This suggests that endothelial cells play a role in the control of neurogenesis in the adult brain, defining the angiogenic niche for neurogenesis. Astrocytes from the adult brain promote the differentiation of adult hippocampal-derived neural progenitor and stem cells toward the neuronal lineage in co-culture, and glial cells are associated with newly generated neuronal cells in the hippocampus [33]. This suggests that glial cells play a role in the control of neurogenesis in the adult brain, by synthesizing factors that promote neuronal differentiation. This defines the astroglial niche for neurogenesis.

Astrocytes from adult hippocampus are very potent in promoting neuronal differentiation of adult hippocampal-derived neural progenitor and stem cells in co-culture, whereas astrocytes from adult spinal cord are ineffective in promoting the differentiation of adult neural progenitor and stem cells [33]. As we reported, adult spinal cord-derived neural progenitor and stem cells transplanted in the adult spinal cord, behave like endogenous progenitor cells, by differentiating in the glial pathway only [7]. The astroglial niche for neurogenesis therefore underlies the differentiation of newly generated neuronal cells in the adult hippocampus and spinal cord. The implication of the angiogenic niche for neurogenesis remains unknown. It suggests a mechanism by which neurogenesis can respond and adapt to the general state of the body. In support of this contention, neurogenesis in the DG and SVZ is increased bilaterally in experimental models of diseases and lesions where only one hemisphere is affected, like in models of epilepsies, after focal cerebral ischemia and traumatic brain injuries [13]. Injury to the CNS would promote transiently the synthesis, and release of trophic factors involved in neurogenesis, like vascular endothelial growth factor and cystatin C [34, 35]. These would reach the neurogenic areas, hippocampus and SVZ, of both hemisphere via blood vessels, and stimulate neurogenesis [13].

NSCs, under certain conditions, give rise to endothelial cells [36]. Hence, not only, the angiogenic niche control NSC activity, but NSCs may contribute to create their own environments in the adult brain and suggest a feed-back regulatory mechanism between NSCs and the angiogenic niche over the control of stem cell activity. The mechanism underlying the feed-back control of NSCs over the angiogenic niche remains to be unraveled and understood. Further, a population of glial cells may correspond to the stem cells in the adult SVZ and DG [8-10]. Hence, populations of astrocytes not only control stem cell activity, but also correspond to NSCs in the adult brain. In all, these data show that neurogenic niches exist in the adult brain, where they control NSC activity. Further, there are relationship between stem cells and their niches in the adult brain.

DEVELOPMENTAL POTENTIAL OF ADULT STEM CELLS

Adult stem cells are multipotent, they generate lineage specific cell types restricted to the tissues from which they are derived. In contrast, pluripotent stem cells generate cells which participate to the three germ layers of the individuals, the ectoderm, mesoderm and endoderm, and the germ cells. Recent evidences show that isolated neural progenitor and stem cells isolated from the adult brain give rise to lineages other than neuronal in vitro and ex vivo, like blood and muscle cells. Conversely, adult stem cells isolated from other issues than the brain, like the skin, blood and bone marrow, give rise to neuronal lineages. Some of these observations originate from cell fusion, transformation or contamination. Others provide strong arguments, by discriminating the phenotype of the cells by X/Y chromosome labeling, to support that adult stem cells may have a broader potential than previously thought; whether such broader potential originates from transdifferentiation or a "true" pluripotentiality remains to be defined [37, 38].

It is hypothesized that stem cell niches may hold the clues of such stem cell plasticity [16, 37]. The host niche would control the fate of stem cells, including of those originating from other tissues. In a new environment, adult stem cells, particularly NSCs, would be instructed by the niche to give lineages of the host tissue. This would involve the regulation, expression or reexpression of molecules that would confer the stem cells a new fate, under a new environment. In support of this contention, niches are conserved structures. They particularly involved developmental signaling pathways, like Notch, bone morphogenetic proteins, Wnt and Sonic hedgehog [16]. These pathways have been recently characterized in adult NSCs [39, 40]. Their regulation would underlie the adoption by adult stem cells of their fate under the control of the microenvironment. In all, these data support a broader potential of adult stem cells, and particularly adult NSCs.

The broader potential of adult stem cells has tremendous implications for cellular therapy, particularly in the CNS. The isolation and culture of stem cells with NSC-capability from other adult tissues than the brain, like the skin, would avoid invasive surgical procedures and their associated risks. They would also permit autologous transplantation, as stem cells, like skin stem cells, could be derived from the own patients. However, the broader potential of adult stem cells remains to be further characterized for the therapeutic potential of adult stem cells to be fully exploited.

THERAPEUTIC IMPLICATION

Contrary to other adult tissues, the CNS does not regenerate after injures. Recent studies reported that new neuronal cells are generated at the sites of degeneration or injury where they replaced some of the lost nerve cells, in the diseased brain and after injury to the CNS, like in Huntington's disease and after experimental strokes. It has been proposed that this corresponds to a regenerative attempt [13]. It is hypothesized that the number of new neurons generated is too low to compensate for the neuronal loss; the number of new neuronal cells correspond to an estimated 0.2% of the degenerated nerve cells in the striatum after focal ischemia [14]. The limited regenerative capacity of the adult brain is yet to be fully understood. The inflammatory response and glial scar, particularly, play a role in preventing and inhibiting cell regrowth and regeneration of the tissue [41, 42]. Nonetheless, the confirmation that neurogenesis occurs in the adult brain and NSCs reside in the adult CNS suggests that the adult brain has the potential for self-repair. Strategies for cellular therapy would involve the stimulation of endogenous progenitor or stem cells or the transplantation of neural progenitor and stem cells to repair the damaged or degenerated pathways [43].

The stimulation of endogenous neural progenitor and stem cells would involve either their stimulation at the sites of degeneration of injury, or to stimulate SVZ neurogenesis to promote the generation of new nerve cells via migration through the RMS to the site of degeneration. With the evidence that niches in the adult brain control stem cell activity, stimulating endogenous neural progenitor and stem cells will require unlocking the molecular and cellular mechanisms inhibiting neurogenesis at the degenerated or injured sites. This includes the inflammatory response and glial scar. Glial cells in the astroglial niche promote neuronal differentiation of neural progenitor and stem cells in vitro, with glial cells from various regions eliciting different potencies on neuronal differentiation [33]. The potency of reactive astrocytes on NSCs would elucidate the role of the glial scar during injuries, and the potential of NSCs for regenerating the injured brain.

Neural progenitor and stem cells have been isolated from human tissues, including postmortem, providing a source of material for cellular therapy [44]. As for the stimulation of endogenous progenitor cells to regenerate pathways, the microenvironment may limit the potential of transplanted neural progenitor and stem cells to achieve functional recovery. In support of this contention, spinal cord-derived neural progenitor and stem cells adopt the fate of endogenous progenitor cells of the spinal cored, by differentiating only into glial cells after transplantation in the spinal cord [27]. The pre-differentiation of neural progenitor and stem cells may offer a mean to overcome the control of the niches over the stem cells. However, pre-differentiated neuronal cells may not integrate and survive in the pre-existing network. In all, this suggests that stem cell therapy in the CNS, either by stimulating of endogenous neural progenitor or stem cells or by transplanting neural progenitor and stem cells, will require the understanding of mechanisms underlying the interaction between stem cells and their niches, and their manipulation to promote stem cell activity and neuronal regeneration.

CONCLUSION

The confirmation that neurogenesis occurs in the adult brain and NSCs reside in the adult NSCs is as important for our understanding of developmental biology, as for cellular therapy. The evidences that neurogenic niches exist in the adult brain further emphasize the relationship between developmental biology, and cellular therapy. Stem cell niches not only control stem cells activity, but also the developmental potential of adult stem cells and therefore their therapeutic potential. Strikingly, NSCs contribute to their environment suggesting a feed-back regulatory mechanism between stem cells and their niches over their activities. In all, the mechanisms underlying stem cell activity and their niches are underlying the developmental and therapeutic potential of NSCs. Future studies will aim at unraveling the molecular and cellular mechanisms of the interactions between NSCs and their niches.

ACKNOWLEDGMENTS

P.T. is supported by grants from the NMRC, BMRC, and the Juvenile Diabetes Research Foundation.

REFERENCES

[1] Gage FH. (2000) Mammalian neural stem cells. *Science.* 287, 1433-8.

[2] Taupin P, Gage FH. (2002) Adult neurogenesis and neural stem cells of the central nervous system in mammals. *J. Neurosci. Res..* 69, 745-9.

[3] Eriksson PS, Perfilieva E, Bjork-Eriksson T, Alborn AM, Nordborg C, Peterson DA, Gage FH. (1998) Neurogenesis in the adult human hippocampus. *Nat. Med.* 4, 1313-7.

[4] Cameron HA, Woolley CS, McEwen BS, Gould E. (1993) Differentiation of newly born neurons and glia in the dentate gyrus of the adult rat. *Neurosci.* 56, 337-44.

[5] Lois C, Alvarez-Buylla A. (1994) Long-distance neuronal migration in the adult mammalian brain. *Science.* 264, 1145-8.

[6] Lois C, Garcia-Verdugo JM, Alvarez-Buylla A. (1996) Chain migration of neuronal precursors. *Science.* 271, 978-81.

[7] Horner PJ, Power AE, Kempermann G, Kuhn HG, Palmer TD, Winkler J, Thal LJ, Gage FH. (2000) Proliferation and differentiation of progenitor cells throughout the intact adult rat spinal cord. *J. Neurosci.* 20, 2218-28.

[8] Doetsch F, Caille I, Lim DA, Garcia-Verdugo JM, Alvarez-Buylla A. (1999) Subventricular zone astrocytes are neural stem cells in the adult mammalian brain. *Cell.* 97, 703-16.

[9] Morshead CM, Reynolds BA, Craig CG, McBurney MW, Staines WA, Morassutti D, Weiss S, van der Kooy D. (1994) Neural stem cells in the adult mammalian forebrain: a relatively quiescent subpopulation of subependymal cells. *Neuron.* 13, 1071-82.

[10] Seri B, Garcia-Verdugo JM, McEwen BS, Alvarez-Buylla A. (2001) Astrocytes give rise to new neurons in the adult mammalian hippocampus. *J. Neurosci.* 21, 7153-60.

[11] van Praag H, Kempermann G, Gage FH. (2000) Neural consequences of environmental enrichment. *Nat. Rev. Neurosci.* 1, 191-8.

[12] Taupin P. (2005) Consideration of adult neurogenesis from basic science to therapy. *Med. Sci. Monit.* 11, LE16-17.

[13] Taupin P. (200) Adult neurogenesis and neuroplasticity. *Restor. Neurol. Neurosci.* 24, 9-15.

[14] Arvidsson A, Collin T, Kirik D, Kokaia Z, Lindvall O. (2002) Neuronal replacement from endogenous precursors in the adult brain after stroke. *Nat. Med.* 8, 963-70.

[15] Jin K, Sun Y, Xie L, Peel A, Mao XO, Batteur S, Greenberg DA. (2003) Directed migration of neuronal precursors into the ischemic cerebral cortex and striatum. *Mol. Cell Neurosci.* 24, 171-89.

[16] Watt FM, Hogan BL. (2000) Out of Eden, stem cells and their niches. *Science.* 287, 1427-30.

[17] Nishimura EK, Jordan SA, Oshima H, Yoshida H, Osawa M, Moriyama M, Jackson IJ, Barrandon Y, Miyachi Y, Nishikawa S. (2002) Dominant role of the niche in melanocyte stem-cell fate determination. *Nature.* 416, 854-60.

[18] Suda T, Arai F, Hirao A. (2005) Hematopoietic stem cells and their niche. *Trends Immunol.* 26, 426-33.

[19] Martin MA, Bhatia M. (2005) Analysis of the human fetal liver hematopoietic microenvironment. *Stem Cells Dev.* 14, 493-504.

[20] Wilson A, Trumpp A. (2006) Bone-marrow haematopoietic-stem-cell niches. *Nat. Rev. Immunol.* 6, 93-106.

[21] Theise ND. (2006) Gastrointestinal stem cells. III. Emergent themes of liver stem cell biology: niche, quiescence, self-renewal, and plasticity. *Am. J. Physiol. Gastrointest Liver Physiol.* 290, G189-93.

[22] Alvarez-Buylla A, Lim DA. (2004) For the long run, maintaining germinal niches in the adult brain. *Neuron.* 41, 683-6.

[23] 23 A. Gritti, E.A. Parati, L. Cova, P. Frolichsthal, R. Galli, E. Wanke, L. Faravelli, D.J. Morassutti, F. Roisen, D.D. Nickel, A.L. Vescovi, (1996) Multipotential stem cells from the adult mouse brain proliferate and self-renew in response to basic fibroblast growth factor. *J. Neurosci.* 16, 1091-100.

[24] Palmer TD, Takahashi J, Gage FH. (1997) The adult rat hippocampus contains primordial neural stem cells. *Mol. Cell Neurosci.* 8, 389-404.

[25] Tropepe V, Coles BL, Chiasson BJ, Horsford DJ, Elia AJ, McInnes RR, van der Kooy D. (2000) Retinal stem cells in the adult mammalian eye. *Science.* 287, 2032-6.

[26] Gage FH, Coates PW, Palmer TD, Kuhn H, Fisher L, Suhonen JO, Peterson DA, Suhr ST, Ray J. (1995) Survival and differentiation of adult neuronal progenitor cells transplanted to the adult brain, *Proc. Natl. Acad. Sci. USA.* 92, 11879-83.

[27] Shihabuddin LS, Horner PJ, Ray J, Gage FH. (2000) Adult spinal cord stem cells generate neurons after transplantation in the adult dentate gyrus. *J. Neurosci.* 20, 8727-35.

[28] Coles BL, Angenieux B, Inoue T, Del Rio-Tsonis K, Spence JR, McInnes RR, Arsenijevic Y, van der Kooy D. (2004) Facile isolation and the characterization of human retinal stem cells. *Proc. Natl. Acad. Sci. USA.* 101, 15772-7.

[29] Qiu G, Seiler MJ, Mui C, Arai S, Aramant RB, Juan E Jr, Sadda S. (2005) Photoreceptor differentiation and integration of retinal progenitor cells transplanted into transgenic rats. *Exp. Eye Res.* 80, 515-25.

[30] Takahashi M, Palmer TD, Takahashi J, Gage FH. (1998) Widespread integration and survival of adult-derived neural progenitor cells in the developing optic retina. *Mol. Cell Neurosci.* 12, 340-8.

[31] Palmer TD, Willhoite AR, Gage FH. (2000) Vascular niche for adult hippocampal neurogenesis. *J. Comp. Neurol.* 425, 479-94.

[32] Shen Q, Goderie SK, Jin L, Karanth N, Sun Y, Abramova N, Vincent P, Pumiglia K, Temple S. (2004) Endothelial cells stimulate self-renewal and expand neurogenesis of neural stem cells. *Science.* 304, 1338-40.

[33] Song H, Stevens CF, Gage FH. (2002) Astroglia induce neurogenesis from adult neural stem cells. *Nature.* 417, 39-44.

[34] Taupin P, Ray J, Fischer WH, Suhr ST, Hakansson K, Grubb A, Gage FH. (2000) FGF-2-responsive neural stem cell proliferation requires CCg, a novel autocrine/paracrine cofactor. *Neuron.* 28, 385-97.

[35] Jin K, Zhu Y, Sun Y, Mao XO, Xie L, Greenberg DA. (2002) Vascular endothelial growth factor (VEGF) stimulates neurogenesis in vitro and in vivo. *Proc. Natl. Acad. Sci. USA.* 99, 11946-50.

[36] Wurmser AE, Nakashima K, Summers RG, Toni N, D'Amour KA, Lie DC, Gage FH. (2004) Cell fusion-independent differentiation of neural stem cells to the endothelial lineage. *Nature.* 430, 350-6.

[37] Taupin P. (2006). Adult neural stem cells, neurogenic niches and cellular therapy. *Stem cell reviews.* In press.

[38] D'Amour KA, Gage FH. (2002) Are somatic stem cells pluripotent or lineage-restricted? *Nat. Med.* 8, 213-4.

[39] Lai K, Kaspar BK, Gage FH, Schaffer DV. (2003) Sonic hedgehog regulates adult neural progenitor proliferation in vitro and in vivo. Nat Neurosci. 6, 21-7. Erratum in: Nat (2003) *Neurosci.* 6, 645.

[40] Lie DC, Colamarino SA, Song HJ, Desire L, Mira H, Consiglio A, Lein ES, Jessberger S, Lansford H, Dearie AR, Gage FH. (2005) Wnt signalling regulates adult hippocampal neurogenesis. *Nature.* 437, 1370-5.

[41] Stoll G, Jander S, Schroeter M. (1998) Inflammation and glial responses in ischemic brain lesions. *Prog. Neurobiol.* 56, 149-71.

[42] Ribotta MG, Menet V, Privat A. (2004) Glial scar and axonal regeneration in the CNS: lessons from GFAP and vimentin transgenic mice. *Acta Neurochir. Suppl.* 89, 87-92.

[43] Taupin P. (2006) Adult neural stem cells and cellular therapy. Journal of Stem cells. 1, 47-55.

[44] Palmer TD, Schwartz PH, Taupin P, Kaspar B, Stein SA, Gage FH. (2001) Cell culture. Progenitor cells from human brain after death. *Nature.* 411, 42-3.

In: Stem Cell Research Progress
Editor: Prasad S. Koka, pp. 59-79

ISBN: 978-1-60456-308-5
© 2008 Nova Science Publishers, Inc.

Chapter 5

AN ANALYSIS OF THE ARGUMENTS UNDERPINNING UK EMBRYONIC STEM CELL LEGISLATION ON THE EMBRYO'S STATUS

Jan Deckers

Institute of Health and Society Faculty of Medical Sciences
University of Newcastle Newcastle-upon-Tyne, NE2 4HH, England, UK

ABSTRACT

With the passing of the Human Fertilization and Embryology (Research Purposes) Regulations 2001, the United Kingdom of Great Britain and Northern Ireland became the first country to pass legislation in support of embryonic stem cell research and research on embryos created by somatic cell nuclear transfer. While the UK legal stance and framework on embryo research are well-known and have attracted significant attention from those with an interest in the ethics of embryo research both in the UK and elsewhere, the arguments on the status of the embryo that were expressed by the members of Parliament and the main advisory bodies involved in this legal debate are less well-known. This article examines the entire range of arguments expressed in support of the Regulations. When the Human Fertilization and Embryology (Research Purposes) Regulations 2001 were passed, the UK had already established a legal framework on embryo research. Since this new legal stance must be understood against the background of these earlier developments and discussions on the status of the embryo, this article will also sketch the legal history of embryo research in the UK, documenting in particular how this legal history has been influenced by the Committee of Inquiry into Human Fertilization and Embryology's arguments on the status of the embryo. While I argue that the validity of all the arguments can be questioned, I argue also that, as long as people have irreconcilable values, no case either for or against granting full moral status to the early embryo can be made that will convince everyone. At the same time, by clarifying my own position on the status of the early embryo, I hope to throw some light on why I believe embryonic stem cell research should not be carried out. Therefore, this article will be of interest to all who have an interest in the ethics of embryonic stem cell research.

Keywords: ethics, UK, legislation, embryo research.

INTRODUCTION

The aim of this article is to provide a critical analysis of the arguments that have been produced on the status of the early human embryo by members of the UK Parliament and its main advisory bodies in the historical development of legislation on embryo research in the UK.[1] The primary focus of my article is to identify and examine the entire range of arguments expressed in support of the Human Fertilisation and Embryology (Research Purposes) Regulations 2001, when the debate was centred mainly on the issue of whether early embryos should be used for embryonic stem cell research and cloning by somatic cell nuclear transfer. While my analysis is restricted to the arguments used in the UK debate, there is no doubt that the arguments discussed here have influenced legal debate elsewhere. As the UK legal framework, because of its relatively early development, has been and is affecting policy debates about embryo research elsewhere, the arguments which underlie this framework are in urgent need of ethical analysis. More generally, this article will be relevant to everyone who has an interest in exploring the ethics of embryonic stem cell research.

EMBRYO RESEARCH BEFORE 1990

For many years, embryo research took place without being regulated. The first major breakthrough in embryo research which gathered unprecedented media attention took place in 1978, when – after many years of research into in vitro fertilisation (IVF) – Louise Brown, the first human baby conceived outside the womb, was born in Oldham Hospital, England. While this event stimulated embryo research in the UK and elsewhere, many UK citizens expressed opposition and demanded legal debate and a clear legal stance.

In 1982 the Department of Health and Social Security of the UK Government decided to set up a Committee of Inquiry into Human Fertilisation and Embryology, chaired by the philosopher Mary Warnock. The Committee came to be known as the 'Warnock Committee'. Apart from a moral philosopher, the Committee included one theologian, two social workers, three legal professionals, and nine people working in a range of medical disciplines. The Committee's remit was to 'examine the social, ethical, and legal implications of recent, and potential developments in the field of human assisted reproduction', and published the 'Warnock Report' in 1984.[1]

Before making its recommendations about research on embryos, the Committee examined the legal status of embryos in the UK, referring mainly to the Offences Against the Person Act 1861 and the Abortion Act 1967. These pieces of legislation make abortion a criminal offence unless any of the grounds specified in the latter Act apply. The Committee therefore concluded that, while 'the human embryo *per se* has no legal status', these statutory provisions do offer some protection.[2] Accordingly, the Committee took the view that, while the embryo should not be granted 'the same status as a living child or an adult' (henceforth:

---

[1] The term 'embryo' will be used to refer to the 'human embryo'; and the term 'early embryo' to refer to an embryo not older than 14 days.

'moral status'), the embryo should be given a 'special status' and be 'afforded some protection in law'.[3] The need for 'protection', however, was not considered to be incompatible with the use of the embryo for research, and therefore with embryo destruction. The Committee thought embryo research can be justifiable, subject to a number of conditions. One condition is what could be called a 'last resort' principle. This is the view that embryo research would be justifiable only when the research could not be done with 'other animals or in some other way' and when they are not 'frivolously or unnecessarily used'.[4] Other conditions are that informed consent is obtained from the gamete donors 'whenever this is possible' and that research is licensed and monitored, which is why the Committee recommended the creation of a new statutory agency to license all embryo research.[5]

The Warnock Report does not provide a clear explanation why the embryo should not be granted moral status. In the introduction to her publication of the Report, Warnock states that the majority of the Committee supported what has become known as the 'gradualist' view, or the view that the embryo's status increases as he or she develops.[6] 'One argument' amongst a 'wide range of opinion' is referred to as 'the strictly utilitarian view' or the view that the question of whether or not to use embryos for research must be settled by 'the balance of benefit over harm, or pleasure over pain'.[7] In this view, 'as long as the embryo is incapable of feeling pain', his or her 'treatment does not weigh in the balance'.[8] Warnock has identified this view as a contributing factor to the majority of her Committee's support for embryo research.[9]

This view is subject to two criticisms. A first problem is how we can know if early embryos are incapable of feeling pain. The Committee refers to the presence of the 'beginnings of the central nervous system'.[10] While we rightly conclude that those who possess a central nervous system normally possess also the capacity to feel pain, this does not rule out the possibility that the capacity to feel pain may be present also where there is no central nervous system. We should be careful not to construe absence of evidence of sentience (the capacity to feel pain) as evidence of absence of sentience. Second, and more importantly, even if we assume that early embryos are incapable of feeling pain, it is not clear why this should matter morally.[2] Anaesthetised children, for example, may not be able to feel pain, but many, if not most people would agree that this does not justify their destruction for research. I suspect that the reason why we believe it is not right to kill children for research, therefore, does not relate to the question if they feel pain, but to the question if it violates their interests in life. By analogy, many people think that it is *prima facie* wrong to cut down trees not because trees might suffer, but because trees have an interest in life. Since it can be argued that early embryos also have an interest in life as – like trees – they sustain themselves and have an orientation towards growth, the *prima facie* wrongness of destroying embryos does not consist in the fact that the act of destruction might cause them pain (as it may not), but in the fact that it destroys their lives. This does not imply that sentience is morally irrelevant. In the unlikely scenario where we are given the choice, *ceteris paribus*, between killing a human being with the capacity to feel pain and killing a human being without this capacity, we should opt to kill the latter. However, this need not imply that the latter has less value than the former. This point can be clarified as follows. Imagine (only for the sake of my argument) being given the choice between scenario A where you are being asked to kill

---

[2] See also Deckers J. (2007, in press) Why Eberl Is Wrong. Reflections on the Beginning of Personhood, *Bioethics* 21 (5).

someone while he or she is anaesthetised, and scenario B where you are being asked to kill that same person without that person being sedated first. While scenario A must be preferred, it is obvious that the reason why this scenario must be preferred does not relate to a difference in value between the person considered for destruction in scenario A and the person considered for destruction in scenario B, given that the same person is being considered as the object of killing in both scenarios. Therefore, while the killing of those who cannot feel pain must be preferred, *ceteris paribus*, to the killing of those who can feel pain, this need not imply that those who lack the ability to feel pain are any less valuable. I conclude, therefore, that this 'argument from sentience' is unconvincing.

Having considered the argument from sentience, the Committee then moves on to recommend that the embryo should not be 'kept alive' or 'used as a research subject beyond fourteen days after fertilisation', excluding 'any time during which the embryo may have been frozen'.[11] The relevant paragraph, however, does not contain a reference to the argument from sentience. Instead, the early embryo is described as 'a potential human being'.[12] Attention is now drawn to 'the formation of the primitive streak', which in an earlier paragraph is identified as 'the latest stage at which identical twins can occur'.[13] The Committee now expresses the view that this is the moment which 'marks the beginning of individual development of the embryo'.[14] No other arguments are provided by the Warnock Committee in support of its gradualist position.

What is problematic about this account is that the view that the early embryo is not a human being, but merely a potential human being, is implausible. Presumably, the Committee arrives at this view on the basis of the view that, as long as the early embryo still has the potentiality to become more than one individual, it cannot be a human individual already. This 'argument from individuality' has had long-lasting appeal in the debate on embryo research in the UK. One of its chief advocates has been the influential Walton of Detchant, a member of the House of Lords, former Dean of Medicine at the University of Newcastle, and past President of the British Medical Association, the Royal Society of Medicine and the General Medical Council. A few years after the publication of the Warnock Report, when embryo research was being discussed in UK Parliament, Walton wrote an article against the views of the Roman-Catholic Cardinal Basil Hume explaining, as a member of the 'Methodist Church', the benefits of embryo research for IVF and the development of pre-implantation genetic diagnosis for the 'prevention' of inherited disease, and claiming that its rejection would be a 'devastating blow ... to that fundamental Christian ethic of aiding those less fortunate than ourselves'.[15] Walton asserted that Hume had 'fallen into the trap of perpetuating several errors of argument and logic which are being regularly advanced by opponents of research', but failed to provide an example to demonstrate why this should be the case.[16] Another Roman-Catholic, however, the Australian Salesian priest Norman Ford, who had just published a book with the title 'When Did I Begin?', was hailed by Walton as 'that eminent (...) scholar'.[17,18] Walton cited Ford and others as 'strongly' supportive of 'the view that individuation of the human embryo (that is, the earliest evidence of the existence of a human individual) cannot be thought to arise until the appearance of the primitive streak', a view which he expressed also in the House of Lords at that time, and again more than a decade later when the House debated the draft Human Fertilisation and Embryology (Research Purposes) Regulations.[19, 20]

The problem with this argument is that the assumption that, as long as something can still become more than one, it cannot be one already, is logically flawed. A flatworm, for example, can divide and reconstitute itself into more than one individual. Yet we do not conclude from the fact that a flatworm can still become more than one individual that it is not an individual already, or that it is only a potential flatworm. This was clear already to Aristotle, who wrote that for plants and insects that can multiply by division, before division, there is 'actually one', but 'potentially many' souls.[21] Therefore, it is not clear why we should conclude that embryos cannot be individuals as long as they can still divide into more than one individual. Since embryologists describe the early embryo as an individual in his or her own right, rather than as part of the female body, describing an early embryo as a potential human being is inappropriate. Without making the assumption that early embryos are human beings, rather than body parts, it seems difficult to explain why they manage to develop in a petri dish, an environment so different from the environment provided by the female body. A plausible view is that they sustain and develop themselves, as human beings in their own right, rather than that external principles direct their development. They are living beings, rather than parts of living beings. Therefore, the Warnock Committee's contention that they merely have a 'potential for life' is flawed. [22]

The Warnock Report's text (as well as Walton's text) could, however, be interpreted in a different way. Even if it were granted that early embryos might be individuals already, it may be the case that what is expressed is the view that what matters morally is whether or not one is an indivisible or irreversible human individual. On this interpretation, early embryos would lack moral status because they might still become more than one human individual. A first problem with the validity of this 'argument from twinning' is that some early embryos might not have the potential to divide and become twins, unless they are forced to divide artificially. Therefore, a distinction in moral value should be made between those early embryos with the potential to divide on their own account and those who lack such potential. In other words, it would not provide support for the view that all early embryos lack moral status. A second, more significant problem is that no argument is provided for why being an indivisible human individual, rather than simply being a human individual, should be what matters morally.

While the Warnock Report's support for embryo research met with strong opposition in UK Parliament, six years went by before the UK Government finally decided on the issue. During this time, support for the recommendations of the Warnock Report increased. The sociologist Michael Mulkay has reported that this may be related partly to the decision, made by the Medical Research Council and the Royal College of Obstetricians and Gynaecologists, to create a Voluntary Licensing Authority (VLA) in 1985, later renamed the Interim Licensing Authority, to regulate embryo research in the absence of statutory provisions. Since the existence of this Authority was mentioned occasionally in Parliament, some members of Parliament might have become convinced that, provided a regulatory authority existed, embryo research was not much of a concern. Mulkay also mentions the influence of 'Progress', a pro-research organisation formed shortly after the creation of the VLA to block the passage of the Unborn Children (Protection) Bill (the 'Powell Bill') which aimed to prohibit embryo research and to make it a criminal offence to fertilise an egg outside the womb unless permission had been given by the Health Secretary and the embryo was destined for implantation.[23] While it would be interesting to examine the arguments on the status of the early embryo expressed by members of UK Parliament in the first major Parliamentary debate on embryo research, such an examination is beyond the scope of this paper.

THE INTRODUCTION OF THE HUMAN FERTILISATION AND EMBRYOLOGY ACT 1990

Eventually, supporters of the the Unborn Children (Protection) Bill were defeated, and six years after the publication of the Warnock Report, the Human Fertilisation and Embryology Act 1990 was passed.[24] In accordance with the recommendations of the Warnock Report, the Act stipulated that the Human Fertilisation and Embryology Authority (HFEA) be established, materialising in 1991, to license and monitor the use of embryos in UK fertilitity clinics and research institutions, and to monitor and review abortion services.[25] Its members are appointed by the Secretary of State for Health, to whom it is accountable. The Act also enshrined the Warnock Report's focus on the primitive streak as the cut-off point: 'a licence cannot authorise ... keeping or using an embryo after the appearance of the primitive streak', where 'the primitive streak is to be taken to have appeared ... not later than the end of the period of 14 days beginning with the day when the gametes are mixed, not counting any time during which the embryo is stored'.[26] By using the words 'not ... keeping or using', the Act avoids referring to the positive acts of destroying or killing which would be required should embryos ever be able to develop beyond that stage outside the womb. The fact that destruction is intended may also be downplayed where the Act states that after the normal statutory storage period of five years cryopreserved embryos 'shall be allowed to perish'.[27][3]

The Warnock Report's concern that early embryos should not be used for trivial research purposes is accommodated by the paragraph that no licence should be granted 'unless the Authority is satisfied that any proposed use of embryos is necessary for the purposes of the research'.[28] The question must be asked if this is consistent with the Act's apparent lack of consideration for embryo destruction resulting from the widespread availability and use of what is erroneously called the combined 'contraceptive' pill and other 'contraceptive' techniques with the potential to block implantation. While the Act amends the 1967 Abortion Act, the use of such 'contraceptive' techniques is exempted from the requirement to satisfy any of the legitimate grounds for pregnancy termination.[29] This is so because the Act stipulates that a woman is not deemed to be 'carrying a child', or to be pregnant, until the embryo has implanted so that a termination of pregnancy cannot occur before implantation.[30] Since embryologists might define conception as the moment when pregnancy starts, this crafty piece of legal tinkering is questionable. This apparent lack of consideration for pre-implantation embryos conceived inside women's bodies appears to sit uncomfortably with the Act's interest in ensuring that any proposed research project on embryos who would normally be conceived outside women's bodies pursues goals deemed sufficiently worthy in relation to reproduction and congenital disease (or – in the words of the Act – is 'necessary or desirable for the purpose of (a) promoting advances in the treatment of infertility, (b) increasing knowledge about the causes of congenital disease, (c) increasing knowledge about the causes of mis-carriages, (d) developing more effective techniques of

---

[3] Incidentally, Walton used similar language whilst discussing PGD, asserting that those embryos 'carrying abnormal genes will simply be allowed to degenerate naturally'. See Walton of Detchant. (1990) Embryo Research – Why the Cardinal is Wrong, Journal of Medical Ethics, 16, 185-186, p. 186. See also Nuffield Council on Bioethics. (2000) Stem Cell Therapy. The Ethical Issues. London. Nuffield Council on Bioethics: par. 21.

contraception, or (e) developing methods for detecting the presence of gene or chromosome abnormalities in embryos before implantation, or for such other purposes as may be specified in regulations).'[31] The apparent inconsistency need not necessarily be understood in terms of a contradiction, though. The Act might have equal regard for all early embryos, irrespective of whether or not they are used for research, but simply regard any destruction of pre-implantation embryos by 'contraceptive' pills to be sufficiently worthwhile, thereby obviating the need to make a distinction between more and less worthy causes in relation to 'contraceptive' technologies.[4]

THE DEBATE ABOUT EMBRYONIC STEM CELL RESEARCH AND CELL NUCLEAR REPLACEMENT

In the late 1990's, however, it became possible to extract embryonic stem cells from early embryos, and many scientists became convinced that these might be used to cure a wide range of disease. Also, research carried out at the Roslin Institute in Edinburgh, which led to the creation of Dolly the sheep, raised the hope that even better therapies might be produced from the combination of embryonic stem cell research and cloning by somatic cell nuclear transfer or cell nuclear replacement. In this context, the UK Government established an Expert Group led by the Chief Medical Officer (including nine people working in medicine or genetics, one person working in veterinary medicine, one ethicist, one specialist in medical law, one theologian, and the Government's Chief Scientific Advisor), charged with assessing the anticipated benefits, risks, and alternatives of new areas of research using embryos, and to advise Government on whether new research purposes should be added to the list approved by the 1990 Act. The Act had authorised the Secretary of State for Health to produce secondary legislation to extend this list.[32] Such an extension had already been recommended in reports issued by two other groups, a report published jointly by the HFEA and the Human Genetics Advisory Commission (the latter provided advice to UK Government on ethical issues related to human genetics before merging with other bodies into the Human Genetics Commission in 1999), and a report by the Nuffield Council on Bioethics (an independent organisation which reports on ethical issues related to biology and medicine).[33,34] The group led by the Chief Medical Officer, Liam Donaldson, then produced a report with the title 'Stem Cell Research: Medical Progress with Responsibility', also known as the 'Donaldson Report'.[35] The Expert Group advocated an extension, and agreed with the gradualist position on the issue of the embryo's status adopted by the Warnock Committee, or the view that the 'respect owed to developing human life is regarded as increasing in proportion to the degree of development of the embryo'.[36] It concluded that embryonic stem cell research may proceed because of its 'great potential to relieve suffering and treat disease'.[37] The recommendations of the Expert Group were then accepted by the UK Government.[38] New legislation was then drafted and debated in both Houses of Parliament. The following arguments – which will be examined shortly – were expressed on

---

[4] The question if embryo destruction can be justified to achieve goals not related to research is addressed in Deckers J. (2007) Why Two Arguments from Probability Fail and One Argument from Thomson's Violinist Succeeds in Justifying Embryo Destruction in Some Situations, *Journal of Medical Ethics* 33 (3) 160-164.

the issue of the status of the early embryo: three arguments from potentiality, an argument from capacities, an argument from probability, an argument from mourning, and an argument from ensoulment.

After the debate, the Human Fertilisation (Research Purposes) Regulations 2001 were made on 24 January 2001, and came into force on 31 January 2001, to allow research (under paragraph 3 of Schedule 2 to the 1990 Act) aimed at '(a) increasing knowledge about the development of embryos, (b) increasing knowledge about serious disease, or (c) enabling any such knowledge to be applied in developing treatments for serious disease'.[39] In this way, the United Kingdom became the first country to legislate on these other types of research, most notably embryonic stem cell research. The scope of acceptable extensions includes the creation of embryos by somatic cell nuclear transfer for research purposes aimed at finding therapies, which has been referred to as 'therapeutic cloning'. While the debate in Parliament was frequently centred on the issue of cloning, Yvette Cooper, Parliamentary Under-Secretary of State for Public Health, tried to reduce some Parliamentarians' concerns about the proposed Human Fertilisation (Research Purposes) Regulations by expressing the view that therapeutic cloning was already legal under the 1990 Act, but that it now could be used for these additional purposes. [40]

The passing of the 2001 Regulations was a blow for Alton of Liverpool, who had pleaded for more time to debate the issues before a decision would be made. Instead, debate was carried on afterwards. In response to a proposal from Walton of Detchant, a House of Lords' Select Committee (henceforth: HL Committee) was appointed to consider and report on the issues connected with stem cell research and human cloning, arising from the new regulations. I had the pleasure of contributing to one of their meetings aimed at gathering 'evidence'.[5] After a series of such meetings, the Committee published a report supporting existing legislation.[41] Unlike the Chief Medical Officer's Expert Group, the HL Committee tried to provide an answer to the question of why early embryos lack moral status. The Report includes the same arguments as those raised in both Houses of Parliament during the debate preceding the new regulations. The validity of these arguments is examined in what follows.

First, there are three arguments from potentiality. The first is the view that the early embryo is potentially human, the second that the early embryo has a passive potentiality to become a human being with moral status, and the third that the early embryo has an active potentiality to become a human being with moral status. The first view, that the early embryo is not a human being, underlies the HL Committee's assertion that even after the embryo has implanted 'there is no trace of human structure'.[42] Likewise, the Donaldson Report stated that the early embryo 'could develop into a human being', or is 'a potential human being'.[43][6] Similar views were expressed in both Houses of Parliament. In the House of

---

[5] The meeting was held at St John's College, University of Durham, 13 November 2001.

[6] In the House of Commons, the same view was expressed by Ian Gibson, Joan Ruddock, and Robert Key (Hansard, 17 November 2000, columns 1192, 1210, 1215), while Michael Fabricant stated that 'a blastocyst is not a human being' (Hansard, 15 December 2000, column 919). The related view, that embryos 'could develop into full human beings', was espoused by Edward Leigh, an opponent of embryo research (Hansard, 31 October 2000, column 629). In the House of Lords, Hunt of Kings Heath, Kennedy of The Shaws, and Taverne referred with approval to the Donaldson Report's claim (Hansard, 22 January 2001, columns 19, 29, 47). In the same House, Sharp of Guildford expressed that the 1990 Act implicitly adopts the view that the early embryo is 'not a human being' (Hansard, 22 January 2001, column 110). Even Alton of Liverpool, a renowned opponent of embryo research, cited the relevant text

Commons, for instance, Yvette Cooper claimed that unneeded embryos 'will not become human beings'.[44] And in the House of Lords, the moral philosopher and member of the Select Committee Onora O'Neill asserted, in her eloquent fashion, that the early embryo 'lacks all internal structure' and that 'there is not even the beginnings of a glimmer of the human form'.[45][7] Yet no explanation is provided for why early embryos should not be human beings, or lack a human structure or form. The fact that early embryos do not possess the organs which foetuses, children or adults have is insufficient to justify this conclusion. Clearly, these people have some idea of what defines as 'the' human structure or form, yet no idea is provided of what this is or why early embryos should be excluded from having such a form. The position that early embryos lack some properties which they ought to possess before they can be considered to be human beings is problematic and has been challenged rightly by Alan Holland. Holland argues that, in a Darwinian world, it makes no sense to stipulate a list of properties which an organism ought to possess in order to be classified to belong to a particular species. Species properties are nothing but contingent manifestations of similar characteristics displayed by a breeding population. A necessary condition for belonging to the species *homo sapiens*, therefore, is not to possess some features which might be similar to the features of others, but to possess the right lineage. Since human embryos are created by other members of *homo sapiens*, they fulfil this condition.[46]

Perhaps these texts bear testimony to a human tendency to use sloppy language: rather than denying the early embryo's humanity as such, the underlying assumption might be that the early embryo is not sufficiently similar to a more developed human being to have moral status. This may be what is behind Evan Harris' claim, expressed in the House of Commons, that an early embryo is 'much smaller than the head of a pin' and that 'there is no question of experimenting on anything that remotely resembles a foetus'; or behind Richard Harries' claim, expressed in a public lecture (Newcastle University, UK, 20 April 2004), that 'to the eye it is a blob of jelly'.[47] In one place, the HL Committee understands this lack of similarity in terms of an alleged lack of (sufficient) identity: because the early embryo still contains cells which develop into the placenta and umbilical cord, and can still divide to constitute identical twins, the view that there is 'such a continuity of identity' as there is between babies and adults 'is less plausible'.[8][48] This is problematic for two reasons. First, neither possession of the capacity to twin nor possession of the capacity to form material which will later be discarded (such as the placenta or umbilical cord) justifies the conclusion that those who possess these capacities lack continuity of identity with those who lack either or both capacities. While two identical twins cannot say that they are identical with each other, it is not problematic for both of them to say that they are identical with the early embryo they once were. As I shall argue further below, this does not, however, lead to the conclusion that for the reverse process of fusion, where two gametes become one embryo,

---

from the Donaldson Report, without comment, in the same House (Hansard, 22 January 2001, columns 63).

[7] See also Howe's views in Hansard, 22 January 2001, column 115.

[8] The same claim was made by Richard Harries in a public lecture (Newcastle University, 20 April 2004). Harries claimed that the possibility of twinning and the fact that the early embryo still contains cells which become the placenta and umbilical cord support the conclusion that it is not possible to 'say, for definite, that was me'. It seems to me that, even if I agreed to the moral importance of having sufficient 'continuity of identity', a stronger conclusion would be needed for its validity, to the effect that it is not possible to 'say, for definite, that was' not 'me'.

there is also a continuity of identity between either gamete and the embryo that forms from them. If we now focus on the fact that early embryos still contain cells which will develop into the placenta and umbilical cord, we must conclude that this fact does not undermine the continuity of identity between early embryos and more developed embryos either. There is no reason why the placenta and umbilical cord should not be regarded as integral parts of the developing embryo. Second, the Committee fails to explain why the existence of (sufficient) 'continuity of identity' should matter morally. If what mattered morally were the fact that one individual was still capable of developing into more than one individual, rather than 'continuity of identity' *per se*, the Committee should have argued that embryos with the capacity to become more than one individual are less valuable compared with embryos who lack this natural capacity, rather than argue that all early embryos lack moral status because some may – without being artificially separated – develop into twins (as I argued earlier on). As mentioned before, even then the question must be asked why the capacity to twin should matter morally. I return to this issue in my discussion of the argument from ensoulment.

The second argument from potentiality is the view that the early embryo has a passive potentiality to develop into a human being with moral status. The Warnock Committee suggested the validity of this view already by writing that some people think that the early embryo is not 'a person' or a 'potential person' because the 'collection of cells ... has no potential for development ... unless it implants', or is 'transferred to a uterus'.[49][9] Likewise, Sally Keeble claimed in the House of Commons that early embryos 'have the potential for life, which is not the same as being alive', while Jenny Tonge claimed that 'human life does not begin until there is sustenance to maintain life from the placenta in the uterus' and that the unimplanted embryo therefore is in 'a pre-life condition'.[50] In the same House, Evan Harris asserted that the blastocyst is a 'disorganised cluster of cells'.[51] The HL Committee provides a more elaborate account, claiming that the view that the early embryo is only a 'potential person' is backed up by 'embryological evidence': 'Although the fertilised egg and blastocyst contain all the genetic signals required for human life, this is true of nearly all cells in the body. ... Although the early embryo contains ... the full genetic potential of any person(s) who may develop from it, it requires many other factors, particularly those provided by the maternal environment in the womb, to enable it to realise that potential.'[52][10] In a meeting of this Committee, Onora O'Neill expressed a similar view, claiming that, in an age of cloning by somatic cell nuclear transfer, this potential was possessed by all cells of the human body.[11] The problem with these claims is that the embryo's potentiality is misconceived. While both somatic cells and embryos require a suitable environment to develop, somatic cells (as well as gametes) need to be transformed into new entities before

---

[9] In the House of Commons, this view was expressed by Ian Gibson, Michael Clark, Evan Harris, and Peter Brand. The last one, for example, expressed the view that 'foetal cells' do not 'have the same status as a unique human being', and that there is a great difference between 'foetal material' and a 'born child' (Hansard, 17 November 2000, columns 1192 and 1194-1195 and Hansard, 19 December 2000, columns 250, 253). In the same House, Howard Stoate was keen to make the point that an embryo created by somatic cell nuclear transfer 'is not a human being', first claiming that such an embryo's lack of potentiality is like the sperm or egg's lack of potentiality, then that this lack of potentiality lies in the fact that 'nobody will be allowed to try' to develop such an embryo (Hansard, 19 December 2000, columns 233-234).

[10] In the House of Lords, Dick Taverne objected to the use of the word 'embryos', preferring 'embryonic cells', for those 'cells which are not to be implanted' (Hansard, 22 January 2001, column 64).

[11] See note 5.

they can acquire the intrinsic potentiality to develop on their own account, while embryos already possess that potentiality. Ann Winterton expressed this view clearly in the House of Commons where she said that the early embryo has 'the capacity to initiate, sustain, control and direct its own development'.[53][12] If the different parts which constitute early embryos were not integrated, it seems hard to explain why these parts develop normally into one or more unified organisms in a continuous process from conception, rather than into a collection of body parts (for example, into a collection of hearts). It goes without saying that embryos need a suitable environment, but that is no different from what adults need to fulfill their potentialities. Embryos necessarily have this intrinsic potentiality for development, irrespective of whether or not their environment is suitable for their development. Neither haploid gametes nor diploid somatic cells possess this intrinsic potentiality.

The third argument from potentiality is the view that merely having the potentiality to reach a certain status is not sufficient to be granted that status already. This argument was adopted by Robert Key in the House of Commons, where he uses the view that there are 'degrees of human sanctity' to support his gradualist account of moral worth.[54] Likewise, the HL Committee supports the view that the early embryo is only 'a potential person rather than ... a person' as follows: 'A medical student is a potential physician, and if he or she qualifies may practise as such; but the potentiality alone does not confer a right to practise. A child is a potential voter but has no claim to be treated as a voter until reaching the age of 18.'[55] By analogy, even if it is granted that the early embryo has an active potentiality to become an adult, this does not mean that he or she has already acquired the status of an adult.[13] The problem with this argument, however, is that the moral status of medical students, children, or 'less saintly people', is no different from the status of physicians, people who are able to vote, or 'more saintly people'. Therefore, while it is true that embryos are potential adults, the argument fails to establish that embryos only have a potentiality to become beings with moral status.

The HL Committee also adopts an argument from capacities to try to justify the view that the early embryo does not deserve a full measure of respect by arguing that 'the basic arguments for respect are focused on (...) beings able to think, act, and communicate'.[56] The problem with this position is that it may rule out others as well, for example human infants. While one could make a case for infant thought, action, and communication, the HL Committee states that the capacity to act - the capacity which one might regard as the most basic amongst the three capacities listed - is lacking in infants. But the HL Committee then argues that respect is 'extended' to infants, which raises the question of why the same measure of respect is not also extended to early embryos.[57] A further problem with this view is that it fails to explain why research is restricted to early embryos, as the argument could be made that these capacities are also lacking in more developed embryos. This suggests that the reason why the HL Committee supports the use of early embryos, but not the destruction of more developed embryos or infants for research, must lie elsewhere. Some

---

[12] See Hansard, 17 November 2000, column 1203 and Hansard, 19 December 2000, columns 243-244.

[13] Richard Harries adopted the same argument in the aforementioned public lecture, using another analogy which has featured in the abortion debate since the early nineteen seventies: the alleged value difference between acorns and oak trees. The validity of this analogy is undermined in Deckers J. (2005) Why Current UK Legislation on Embryo Research Is Immoral. How the Argument from Lack of Qualities and the Argument from Potentiality Have Been Applied and Why They Should Be Rejected, *Bioethics* 19 (3) 251-271, p. 264.

support for the argument from sentience seems to be given where the HL Committee writes that the fourteen days limit for research 'has an objective justification insofar as it represents the stage at which the primitive streak … begins to appear', a stage before which (as clarified in a preceding paragraph) 'there can be no sentience'.[58][14] While it may be an objective fact that the primitive streak appears around the fourteenth day, the HL Committee does not explain why this should be an 'objective justification' for research on embryos younger than fourteen days. As I shall argue in the final section, I am not convinced that the existence of variable capacities provides a moral justification for discrimination amongst humans.

A fifth argument bases the view that early embryos have little value on the view that there is a high probability that they may not be able to survive beyond a certain point regarded as critical, for example implantation. In the House of Commons, this argument was supported by Evan Harris and Peter Brand.[59] Harris, who acknowledges that he owes this view to 'Church of England theologians', appears to be bothered by the possibility that the 'majority of stars in heaven' might be embryos who never implanted, which he feels is easier to reconcile with the idea that they lack moral status than with the idea that they have such status. The HL Committee appears to approve of this argument as well. Having claimed that the natural loss rate of early embryonic death is 'as high as 75 per cent', the Committee describes the following information as 'consistent with' its 'gradualist view': 'Although would-be parents may feel sad at the natural loss of early embryos before implantation, there is no public mourning ritual associated with it, nor is there for the loss of surplus embryos left over from IVF treatment.'[60][15] The use of the word 'loss' to describe the destruction of spare embryos downplays the intentional act of destruction and suggests that 'the loss of surplus embryos' would be no different from the 'natural loss of early embryos'. While the Committee does not state explicitly that the high probability that embryos decay naturally counts as a reason for their lack of moral status, this may be what is suggested here. Whether or not this is a valid interpretation, at least the Committee's chairman, Richard Harries, used the argument in the aforementioned public lecture. It was also used by Robert Winston, a specialist in reproductive medicine, in the House of Lords' debate. Winston claimed that 'twelve per cent of human embryos in an in vitro fertilisation programme implant' and 'something probably similar happens in nature'. This would support the view that, 'by nature's standards', the human embryo is 'not sacrosanct'.[61] The argument is problematic. First, it sums over the survival odds of each individual embryo. While it may be true that '75 per cent' (or 88 per cent) of early embryos die young, this does not establish that each embryo has a '75 per cent' (or 88 per cent) chance of dying early. For some, the chances of implanting or being born might be much higher than for others, who might be doomed from the start. Therefore, if the probability of a successful outcome (whether this be implantation or birth) would be what mattered morally, the argument should not be that all early embryos have little value, but only those who have a relatively small chance of a successful outcome. A second problem with the argument, however, is that many people would agree that, when it comes to assessing the moral worth of children or adults, the question of how large or small

---

[14] In the House of Commons, the argument was endorsed by Evan Harris, Michael Fabricant, and Yvette Cooper (Hansard, 31 October 2000, column 627; Hansard, 15 December 2000, column 919; Hansard, 19 December 2000, column 212).

[15] The argument is also supported by Onora O'Neill, a moral philosopher and member of the House of Lords (Hansard, 22 January 2001, column 68).

their chances are of surviving up to a certain stage are irrelevant. Therefore, the view that there is a high probability that early embryos might die young fails to justify that they lack moral status.

The above quotation from the HL Committee also contains an argument from mourning, supported also by Richard Harries in the aforementioned public lecture, and by Robert Key in the House of Commons: the absence of 'public mourning ritual' would support the view that the early embryo lacks moral status.[62] The problem with this argument is that we do not use the fact that the deaths of some children or adults do not cause a lot of grief in other people as evidence for their lack of moral status. Therefore, it is not clear why it should support the view that early embryos lack moral status. One could object that there are other factors which may account for why some born people's deaths are not mourned over (much), for example that it depends on a relative lack of friends or close friends, but that the death of anyone belonging to the class of early embryos is not mourned over much. The counterobjection is that this is not true, since many people go through severe mourning experiences when they experience early miscarriages. Another reason why there might be no 'public mourning ritual' for the death of an early embryo is that many women may not know that they are pregnant. Whether or not they are, it is surely the case that early embryos are not 'public' persons in the sense that they are not visibly present as independent persons within society, which may account for the lack of 'public mourning ritual'. Finally, not too long ago many slaves' deaths may not have been accompanied by a 'public mourning ritual'. Yet I hope not many people today would argue that this should count as evidence for their lack of moral status. Even if it were the case that embryos' deaths are not mourned over much, it is a naturalistic fallacy to conclude from this that embryos' deaths ought not to be mourned over much, or that moral status ought not be granted to them. I conclude, therefore, that the argument from mourning is unconvincing.

Finally, two versions of the argument from ensoulment, or the view that early embryos are not ensouled and therefore lack moral status, have been supported in this debate. The chief advocate of the first version - identified also as one out of two Christian views by the HL Committee - has been the HL Committee's chairman, Richard Harries.[63][16] Harries refers to Aristotle's embryology, where 'there is first a vegetable soul, then an animal soul', and only at 40 days after conception for men and at 90 days for women, an intellective soul.[64,65] While it is correct that Aristotle wrote that the fetus starts to move and differentiate only around 40 days after conception for boys and around 90 days for girls, Harries understands that this implies that, for Aristotle, 'it is only at the last point that there is, properly speaking, a human being'.[66] He claims that a similar position was adopted in the Christian Church, at least in the 'western' tradition, until the nineteenth century when the view that ensoulment occurs at conception became 'firmed up'.[67] Harries claims that his gradualist account is supported by 'this understanding of the western tradition'.[68] This view is subject to a number of criticisms.

First, Harries claims that the present Catholic position that ensoulment occurs at conception stems largely from the nineteenth century, and refers to a number of early and

---

[16] The argument was also used by Evan Harris in the House of Commons' debate. Harris refers to a paper written by the Anglican theologian John Polkinghorne (with similar views as those held by Harries regarding ensoulment) (Hansard, 19 December 2000, column 253). Polkinghorne's paper can be found at http://www.cofe.anglican.org/info/socialpublic/cnr.html

medieval Christian writers who thought otherwise. While Harries is correct that some Christians have supported 'delayed ensoulment' theories, in his article on 'The Appeal to the Christian Tradition in the Debate about Embryonic Stem Cell Research', David Jones has illustrated that the view that ensoulment occurs at conception was by no means absent from the early Christian tradition.[69, 70]

Second, Harries claims that the delayed ensoulment theory finds Biblical support in Exodus 21.22-24, where a distinction is made between a 'fully formed' and a 'not fully formed' fetus (literally: 'μή ἐξεικονισμένον', or 'not yet so formed as to be a copy or portrayal of the human form'), and where anyone found involved in the destruction of the latter is held to be subject to less severe penalties compared with anyone found guilty of destruction of the former.[71] In fact, the distinction is absent from the Hebrew text, but stems from a mistranslation in the Septuagint, the influential pre-Christian Greek translation of the Scriptures. In the other passages where the Hebrew word 'אָסוֹן' occurs (Genesis 42.4,38;44.29), the word could not possibly be translated by 'fully formed'.[17] In fact, the Septuagint translates the word in these other passages as 'sickness'. In the context of Exodus 21.22-24, the best translation may be 'harm' or 'injury'. In other words, the gradation in penalty applies not to the issue of whether or not the fetus is 'fully formed' (which, anyhow, need not be synonymous with 'ensouled'), but to the question if the pregnant woman suffers further injury apart from having a miscarriage. Harries is right, though, that the distinction between 'formed' and 'unformed' then found its way into Christian theology through Augustine (as documented by Jones), who made reference to the distinction relying on an old Latin translation of the Septuagint.[72,73] Also, Jones has documented that some writers then started to identify the distinction between 'fully formed' and 'not fully formed' with the distinction between animated and inanimated, providing the example of an anonymous work from the fifth century, originally thought to be Augustine's.[74, 75]

Third, the suggestion that, for Aristotle, the intellective soul was only present a long time after conception is flawed. Aristotle thought the first movements could be registered on the fortieth and the ninetieth day after conception for boys and girls respectively, but doubted the accuracy of his conclusions. More importantly, he held that the early embryo contained all three souls from the beginning, not in act, but in potency. While the vegetative power would be actualised before the sensitive power, and the sensitive power before the intellective power, all three powers were held to be present before their actualisation, as different parts of one soul.[76, 77]

Finally, and most problematically, for Aristotle, the cause of the creation of these three successive stages of the soul was the sperm, while for Aquinas (the first Christian theologian to use Aristotle's work extensively) the sperm was the cause of the creation of the 'lower' souls, yet not of the intellective soul (which he thought had to be created directly by God, that is: without the use of sperm), and therefore the sperm had to continue existing until this process was completed.[78,79] Yet, since Karl Ernst von Baer observed and described the mammaliam ovum for the first time in 1827, complementing Reinier de Graaf's first observations of the ovarian follicle in 1672 and Antoni van Leeuwenhoek's first observations of the sperm in 1677, it was only a matter of time before further observations would prove beyond reasonable doubt that both were involved in the creation of an embryo.[80] Important breakthroughs in the latter part of the nineteenth century were the work of Gregor Mendel,

---

[17] I am grateful to Robin Salters for the help I received to examine this issue.

who discovered from his observations of pea plants that organisms inherit two genes for each trait, one from each parent, and the observation (originally of a nematode worm) and description of meiosis during gametogenesis in animals by Edouard Van Beneden.[81, 82] The latter observation corroborated Wilhelm August Oscar Hertwig's conclusion, based on the observation of sea urchins, that sperm and egg fuse in animals that reproduce sexually.[83] Thus, we now know that the sperm does not continue to exist after conception. This is incompatible with the view that the sperm might be the cause of ensoulment at any stage after conception. The idea that the early embryo is alive solely by virtue of the soul of the male parent, present in the sperm, and that the early embryo is not an independent living being until some time after conception, flies in the face of modern embryology. In the light of these developments, it seems strange how anyone could still hold on to the view that ensoulment does not occur at conception. Even more bizarre is Walton of Detchant's statement, which he claims (erroneously) was presented as Aquinas's view 'until the middle of the 19th century', that 'the foetus' does 'not develop as an independent human being and that life' does 'not begin until the foetus' is 'capable of independent existence outside the womb'.[84] This great progression in embryological understanding achieved over the course of the nineteenth century, rather than - as Harries claims - the greater incidence of abortion resulting from greater technical expertise in the carrying out of abortions, may also have prompted pope Pius IX into deciding, in 1869, that equal penalties should apply to early and late abortions.[85,86]

I shall now turn to the second version of the argument from ensoulment, a variant of the argument from individuality found in the Warnock Report. It is the view that early embryos cannot be ensouled because they might still divide into more than one organism or fuse to become one organism, and that they therefore lack moral status. Peter Brand expressed this argument in the House of Commons, while Richard Harries has claimed that the HL Select Committee held that the possibility of twinning excludes the early embryo from being a person, and therefore presumably from being an ensouled human being.[87,88] If we understand the soul in the traditional (Aristotelian) sense as the principle that provides organisation and direction to the human body, the implication is that early embryos lack such a principle. In other words, early embryos would either be like body parts which might have relative autonomy but which are ultimately controlled by the souls of the bodies in which they are situated, or like machines which are controlled entirely by external forces. Yet, since early embryos appear to develop autonomously, and given the fact that a strong continuity exists between a fourteen and a fifteen day old embryo, irrespective of whether or not twinning or fusion still occurs on the fourteenth day, the view that the principle which directs the development of early embryos is different from the principle which directs the development of more developed embryos is implausible. A more plausible view is that ensoulment occurs at conception, given that no such continuity exists between gametes kept in separation from each other and the early embryo who comes into existence when gametes are brought together. Another plausible view is that the cause of ensoulment, the principle which fuses sperm and egg, is not an external agent, but the early embryo who creates himself or herself out of the materials provided by the sperm and egg. The Christian theological perspective supported by Harries and many others, where an external agent, God, intervenes to infuse the soul, lacks a satisfactory answer to the problem of evil. What it cannot answer is, for instance, the question why a good God would infuse souls inside the bodies of women after rape? That Harries wants to hold on to a belief in God's goodness seems apparent from what has been reported to be his concern that, if the view that ensoulment occurs at conception were accepted, together with the view that three quarters of all embryos fail to implant, 'three quarters of heaven would be populated by souls that lived for less than a week'.[18] The reason why Harries would find this conclusion 'strange', and therefore the first premise unacceptable, may relate to the view that his belief in the goodness of God is incompatible with the idea that many souls die at such an early stage.[89] Regardless of this theological critique, what is more problematic is that those who base the view that the early embryo is not ensouled either on Harries' concern over the possibility of twinning or on Brand's concern over the possibility of fusion, are unwilling to change traditional embryological conceptions in the light of modern embryology. Before the advent of modern embryology, the phenomena of twinning and embryonic fusion were understood even less than they are today, and therefore there was no need to conceive of the possibility of twinning or fusion occurring after conception. Accordingly, it was thought that a soul could not divide into souls or fuse with other souls. This view fitted common sense, as common observation was sufficient to conclude that born humans no longer twin or fuse with other humans to become new human individuals. Yet, since we have now managed to observe what takes places in the early stages after conception and know that twinning and fusion can occur after conception, why should we hold on to the view that a soul cannot divide into souls or fuse with another soul? Why should either the splitting of one soul into two or more souls, or the merging of multiple souls into one soul, be regarded as impossible? I conclude, therefore, that both versions of the argument from ensoulment are flawed.

SHOULD EARLY EMBRYOS BE GRANTED MORAL STATUS?

In this article, an overview of the arguments on the status of the early embryo that have been used by the main legal advisory bodies and members of Parliament in support of legal developments related to embryo research in the UK has been provided, the main focus being on the most recent developments. I have shown that none of these arguments are convincing. While the UK is currently regarded as being one of the most 'liberal' countries as far as the legitimacy and scope of embryo research is concerned, many have expressed the view that its legal stance and legislative framework can be an inspiration for other legislatures.[19] What my paper has shown is that the arguments underpinning this framework provide a shaky, rather than - as claimed by the House of Commons' Science and Technology Select Committee on Human Reproductive Technologies and the Law - a 'firm foundation' for legislation.[90] However, while it is one thing to undermine a range of arguments that have been used to deny moral status to the early embryo, it is another matter to make a convincing case for why the early embryo should be granted such status. On this issue, I agree with David Hume's position, cited with approval by Warnock, that morality is 'more properly felt than judg'd of'.[91,92] Warnock rightly adds, though, that the fact 'that a decision is based on sentiment

---

[18] Robert Key claims that these are Harries' words, with approval of his view (Hansard, 17 November 2000, column 1215). Evidence for the rightness of Key's claim is Harries R. (2004) Why Limited Cloning Is Right and Necessary, *Church Times* (20 August).

[19] See for example Herder M. (2002) The UK Model: Setting the Standard for Embryonic Stem Cell Research?, *Health Law Review*, 10, 14-24.

by no means entails that arguments cannot be adduced to support it'.[93] These arguments, however, might not convince those who have different values or sentiments. This need not result in a kind of moral paralysis where those with irreconcilable values refuse to engage in debate with one another, or disparage the positions held by others by labelling them as irrational. Neither is it necessary to believe that those with irreconcilable values will never be able to reconcile their differences. While it is beyond the remit of this paper to make an elaborate case against the gradualist account of the embryo's status which has been supported by these legal developments, my final paragraph aims to sketch the position I have developed elsewhere, that early embryos should be granted moral status. I also illustrate how this position might be undermined both by an objectifying discourse and by the view that assigning moral status to the embryo is necessarily incompatible with embryo destruction. [94]

My view is that we should promote equality between all human beings and treat all humans as equal, rather than assign different values to different human beings depending on how many properties or capacities each human being may possess. If discrimination against infants or children, even though they are a long way from being adults, is not acceptable, then discrimination against early embryos is not acceptable either. While many people may not support this analogy, a number of people expressed their appreciation for such a position when I talked about the subject at a recent symposium in Oxford.[20] Discourse about the human embryo often portrays the embryo as if he or she were an abstract, alien entity, the product of those who experiment with substances in test tubes in laboratories. The moral position that human embryos are suitable objects for research might be favoured by this kind of discourse. When I showed the first ultrasound scan of my daughter, no more than eight weeks after her conception, followed by a picture of my six year old daughter now, and asked rhetorically if adopting a gradualist position might mean accepting the view that she was quite worthless around the time her first picture was taken, many conference attendees realised that a different discourse was possible, and started questioning the gradualist account, reflected in the reactions I received after my presentation. Another reason why many people may reject the view that the early embryo has moral status might relate to the widely held view - adopted also by the HL Committee - that such a position is incompatible with embryo destruction.[95] The view is then rejected because the position that embryo destruction should not be allowed under any circumstances, including, for instance, cases of pregnancy after rape, is regarded as unacceptable. I have argued elsewhere, however, that the view that ascribing moral status to the embryo is incompatible with all forms of embryo destruction, is flawed, but that it is incompatible with the destruction of embryos for research.[96] Killing a human being is a serious act of violence, irrespective of whether or not embryos feel pain. Children, who have not had the same amount of exposure to the 'laboratory discourse' on embryos, are no less horrified when they hear about embryo research by being told that these early embryos are so small and undeveloped that they do not even feel that they are being destroyed. What may be more acute in their understanding is that human embryos are not aliens, but that – to use the words used by Edward Leigh expressed in the House of Commons – 'we were all at that stage once'.[97] Perhaps the paediatrician John Wyatt is right when he claims, in his article on 'Medical Paternalism and the Fetus', that 'nearly all of us share deep-rooted intuitions that the

---

[20] The symposium on Bioethical Issues at the Beginning, Middle, and End of Life, was organised by the Oxford Forum for Medical Humanities, 2 June 2006, St Anne's College, Oxford.

protection, support, and nurturing of vulnerable human beings ... is an essential duty of a civilised society.'[98] While supporters of embryonic stem cell research might claim that the protection of vulnerable people is precisely what their research is about, the means by which this must be achieved should not include the destruction of other people. However, in cultures where many nonhuman animals are objectified and killed gratuitously, the killing of human embryos for less trivial purposes may not be surprising. From the perspective of Ken Wilber's transpersonal psychology, the destruction of others in what Wilber calls 'Atman projects' is a coping strategy used by individuals anxious about their mortality.[99] From this perspective, the words of Alison Murdoch - a member of the Newcastle Fertility Centre which was granted the first UK licence in 2004 to create embryos by somatic cell nuclear transfer - who is reported to have said that spare embryos 'have no more moral status than blood taken from a patient', carry deeper meaning.[100][21] While religious sacrifices of blood may have become rare events in the modern world, the destruction of embryos for research might fulfil a similar psychological function: to provide an illusory sense of security in the fight against one's own mortality and one's ultimate lack of control.

REFERENCES

[1]   Warnock M. (1985) *A Question of Life*. The Warnock Report on Human Fertilisation and Embryology, Oxford, Blackwell, vi.
[2]   Warnock M. *A Question of Life*, par. 11.16.
[3]   Warnock M. *A Question of Life*, par. 11.17.
[4]   Warnock M. *A Question of Life*, par. 11.17 and 11.18.
[5]   Warnock M. *A Question of Life*, par. 11.18, 11.24, and 13.3.
[6]   Warnock M. A Question of Life, xiv.
[7]   Warnock M. *A Question of Life*, par. 11.20.
[8]   Warnock M. *A Question of Life*, par. 11.20.
[9]   Warnock M. A Question of Life, xv.
[10]   Warnock M. *A Question of Life*, par. 11.20.
[11]   Warnock M. *A Question of Life*, par. 11.22.
[12]   Warnock M. *A Question of Life*, par. 11.22.
[13]   Warnock M. *A Question of Life*, par. 11.5 and 11.22.
[14]   Warnock M. *A Question of Life*, par. 11.22.
[15]   Walton of Detchant. (1990) Embryo Research – Why the Cardinal is Wrong, *Journal of Medical Ethics*, 16, 185-186.
[16]   Walton of Detchant. Embryo Research – Why the Cardinal is Wrong, p. 186.
[17]   Walton of Detchant. Embryo Research – Why the Cardinal is Wrong, p. 185.
[18]   Ford N. (1988) When Did I Begin? Conception of the Human Individual in History, Philosophy, and Science, Cambridge, University Press.
[19]   Walton of Detchant. Embryo Research – Why the Cardinal is Wrong, p. 185.
[20]   Hansard, 22 January 2001, column 104.
[21]   Aristotle. On the Soul, I, chapter 4 and II, chapter 2.

---

[21] Barnett A. and McKie R. (2004) UK to Clone Human Cells, *The Observer* (13 June).

[22] Warnock M. *A Question of Life*, par. 11.28.

[23] Mulkay M. (1997) The Embryo Research Debate. *Science and the Politics of Reproduction*, Cambridge, University Press, p. 41, 58, 63.

[24] UK Parliament. (1990) Human Fertilisation and Embryology Act 1990, London, HMSO.

[25] UK Parliament. (1990) *HFE Act 1990*, sections 5-10.

[26] UK Parliament. (1990) *HFE Act 1990*, section 3 par. 3-4.

[27] UK Parliament. (1990) *HFE Act 1990*, section 14 par. 1c.

[28] UK Parliament. (1990) *HFE Act 1990*, schedule 2, par. 3.6.

[29] UK Parliament. (1990) *HFE Act 1990*, section 37.

[30] UK Parliament. (1990) *HFE Act 1990*, section 2 par. 3.

[31] UK Parliament. (1990) *HFE Act 1990*, schedule 2, par. 3.2.

[32] UK Parliament. (1990) *HFE Act 1990*, section 45 and schedule 2, par. 3.2 and 3.3.

[33] Human Genetics Advisory Commission and Human Fertilisation and Embryology Authority. (1998) *Cloning: Issues in Reproduction, Science and Medicine,* London, HGAC/HFEA.

[34] Nuffield Council on Bioethics. (2000) *Stem Cell Research*. The Ethical Issues, Nuffield Council on Bioethics, London.

[35] Chief Medical Officer's Expert Group (Reviewing the Potential of Developments in Stem Cell Research and Cell Nuclear Replacement to Benefit Human Health). (2000) *Stem Cell Research*: *Medical Progress with Responsibility,* London, Department of Health.

[36] Chief Medical Officer's Expert Group. *Stem Cell Research,* section 4.2.

[37] Chief Medical Officer's Expert Group. *Stem Cell Research,* executive summary: conclusions 24-25 and recommendation 1.

[38] Department of Health. (2000) Government Response to the Recommendations Made in the Chief Medical Officer's Expert Group Report '*Stem Cell Research: Medical Progress with Responsibility*', Norwich, HMSO.

[39] Statutory Instrument 2001 No. 188. *The Human Fertilisation and Embryology* (Research Purposes) Regulations 2001, London, HMSO.

[40] Hansard, 17 November 2000, column 1181.

[41] House of Lords. (2002) *Stem Cell Research*. Report From the Select Committee, Published by Authority of the House of Lords, London, HMSO, section 4.21 and recommendation 7.

[42] House of Lords. *Stem Cell Research*, section 4.2.(d).

[43] Chief Medical Officer's Expert Group. *Stem Cell Research*, section 4.17 and executive conclusion 26.

[44] Hansard, 15 December 2000, column 885.

[45] Hansard, 22 January 2001, column 67.

[46] A. Holland. (1990) A Fortnight of My Life is Missing: A Discussion of the Status of the Human 'Pre-Embryo', *Journal of Applied Philosophy* 7 (1990) 25-37.

[47] Hansard, 31 October 2000, column 627.

[48] House of Lords. *Stem Cell Research*, section 4.11.

[49] Warnock M. *A Question of Life*, par. 11.15 and 11.28.

[50] Hansard, 19 December 2000, column 256 and Hansard, 15 December 2000, column 887.

[51] Hansard, 15 December 2000, column 889.

[52] House of Lords. *Stem Cell Research*, sections 4.11 and 4.12.

[53] Hansard, 17 November 2000, column 1203 and Hansard, 19 December 2000, columns 243-244.

[54] Hansard, 19 December 2000, column 239.

[55] House of Lords. *Stem Cell Research*, sections 4.10 and 4.11.

[56] House of Lords. *Stem Cell Research*, section 4.7.

[57] House of Lords. *Stem Cell Research*, section 4.7.

[58] House of Lords. *Stem Cell Research*, sections 4.22 and 4.2 (d).

[59] Hansard, 15 December 2000, column 886 and Hansard, 17 November 2000, column 1195.

[60] [60] House of Lords. *Stem Cell Research*, sections 4.2.(d) and 4.13.

[61] Hansard, 22 January 2001, column 100.

[62] Hansard, 17 November 2000, column 1215 and Hansard, 19 December 2000, column 239.

[63] House of Lords. *Stem Cell Research*, appendix 4.

[64] Hansard, 22 January 2001, column 36.

[65] Aristotle. On the History of Animals, VII, ch. 3.

[66] Hansard, 22 January 2001, column 36.

[67] Hansard, 22 January 2001, columns 35, 37.

[68] Hansard, 22 January 2001, column 37.

[69] Jones D. (2005) The Appeal to the Christian Tradition in the Debate about Embryonic Stem Cell Research, *Islam and Christian-Muslim Relations* 16 (3), 265-283.

[70] Jones D. (2005) The Human Embryo in the Christian Tradition: A Reconsideration, *Journal of Medical* Ethics 31, 710-714.

[71] Hansard, 22 January 2001, column 36.

[72] Jones D. (2005) The Appeal to the Christian Tradition in the Debate about Embryonic Stem Cell Research, p. 273.

[73] Augustine. (1845) Questions on Exodus 80, in J.P. Migne (ed.), *Patrologia Latina*, Paris, Migne, vol. 34, col. 626.

[74] Jones D. (2005) The Appeal to the Christian Tradition in the Debate about Embryonic Stem Cell Research, p. 274.

[75] Anonymous. (1845) Quaestiones in Veteri Testamento XXIII, in J.P. Migne (ed.), *Patrologia Latina*, Paris, Migne, vol. 35, col. 2229.

[76] Aristotle. *On the Soul*, II, chapter 1.

[77] Aristotle. *On the History of Animals*, VII, chapter 3.

[78] Aristotle. On the Generation of Animals, II.

[79] Aquinas. (1963-1975) Summa Theologiae, Cambridge, Blackfriars edition with translation, Ia, q.118, a.2, ad2.

[80] Von Baer K.E. (1827) *De Ovi Mammalium et hominis genesi,* Leipzig, Voss.

[81] Mendel G. (1865) Versuche über Pflanzenhybriden, in *Verh. Naturforsch. Ver. Brünn* 4, 3-47.

[82] Van Beneden E. (1883) Recherches sur la maturation de l'œuf et la fécondation. Ascaris Megalocephala, in *Arch. Biol.* 4, 610-620.

[83] Hertwig O. (1876) Beitrage zur Erkenntnis der Bildung, Befruchtung, und Theilung des thierischen Eies, *Morphol. Jahr.* 1, 347-452.

[84] Hansard, 22 January 2001, column 104.

[85] Hansard, 22 January 2001, column 36.

[86] Pius IX. (1869) Constitution *Apostolicae Sedis*, in Pius IX. (1971) Pii IX Pontificis Maxima Acta, Graz, Akademische Druck- und Verlagsanstelt, 305-331.

[87] Hansard, 17 November 2000, column 1195.

[88] Harries R. (2005) Delivering Public Policy. The Status of the Embryo and Tissue Typing, *Studies in Christian Ethics* 18 (1) 57-74, p. 63.

[89] Harries R. (2004) Why Limited Cloning Is Right and Necessary, *Church Times* (20 August).

[90] House of Commons' *Science and Technology Select Committee on Human Reproductive Technologies and the Law.* (2005) Fifth Report of Session 2004-2005, London, HMSO, volume 1, chapter 3, par. 28.

[91] Hume D. (1978) *A Treatise of Human Nature*, Second edition with text revised and notes by P.H. Nidditch, Oxford, Clarendon Press, p. 470.

[92] Warnock M. A Question of Life, x.

[93] Warnock M. A Question of Life, x.

[94] Deckers J. (2005) Why Current UK Legislation on Embryo Research Is Immoral. How the Argument from Lack of Qualities and the Argument from Potentiality Have Been Applied and Why They Should Be Rejected, *Bioethics* 19 (3) 251-271.

[95] House of Lords. *Stem Cell Research*, section 4.23.

[96] Deckers J. (2007) Why Two Arguments from Probability Fail and One Argument from Thomson's Violinist Succeeds in Justifying Embryo Destruction in Some Situations, *Journal of Medical Ethics* 33 (3) 160-164.

[97] Hansard, 31 October 2000, column 629.

[98] Wyatt J. (2001) Medical Paternalism and the Fetus. *Journal of Medical Ethics* 27 (suppl II) ii5-ii20, p. ii17.

[99] Wilber K. (1983) Up from Eden. A Transpersonal View of Human Evolution, London, Routledge and Kegan Paul.

[100] Barnett A. and McKie R. (2004) UK to Clone Human Cells. *The Observer* (13 June).

In: Stem Cell Research Progress
Editor: Prasad S. Koka, pp. 81-94

ISBN: 978-1-60456-308-5
© 2008 Nova Science Publishers, Inc.

Chapter 6

A TETRACYCLINE-INDUCIBLE LENTIVECTOR SYSTEM BASED ON EF1-α PROMOTER AND NATIVE TETRACYCLINE REPRESSOR ALLOWS IN VIVO GENE INDUCTION IN IMPLANTED ES CELLS

David M. Suter[*1], *Xavier Montet*[2], *Diderik Tirefort*[1], *Laetitia Cartier*[1] *and Karl-Heinz Krause*[1]

[1] Department of Pathology and Immunology,
University of Geneva Medical School, Switzerland
[2] Department of Cellular Physiology and Metabolism,
University of Geneva Medical School, Switzerland

ABSTRACT

Inducible transgene expression remains an important challenge in embryonic stem cell research. It is desirable because it allows to control the level and timing of gene expression in stem cells. Here we describe a tetracycline-regulatable system based on the repression of a modified EF1-α promoter (EF1-αTetO2) by the native tetracycline repressor. EF1-αTetO2 activity was low in tetracycline repressor-expressing cell lines, but was strongly enhanced upon addition of tetracycline. Effective transgene expression was achieved within 24 hours of tetracycline addition; expression levels returned to baseline within 24 hours after tetracycline withdrawal. Graded transgene expression levels were achieved by increasing tetracycline concentrations. Efficient induction was observed in undifferentiated ES cells as well as ES cell-derived neurons. Upon subcutaneous cell implantation in mice, the system allowed in vivo induction of

---

[*] Correspondence: David Suter, Department of Pathology and Immunology, University of Geneva Medical School, 1 rue Michel-Servet, CH-1211, Geneva, Switzerland. Telephone: + 41223794141; Fax: + 41223794132; E-mail:David.Suter@medeci ne.unige.ch

transgene expression. To our understanding, this is the first description of a switch-on expression system allowing transgene induction in implanted ES cells.

INTRODUCTION

Genetic engineering is a major challenge of embryonic stem (ES) cell research, with applications ranging from developmental studies to future cell therapy. Pharmacologically-regulated systems are of particular interest, providing a useful tool for cell differentiation studies and offering attractive possibilities for experimental cell therapy. Various systems allowing regulated transgene expression are currently available. Among these, tetracycline-controlled expression remains the most widely used option and was shown to be effective in vitro and in vivo [1, 2]. In addition, tetracyclines are well tolerated in humans and therefore ideally suited for a potential clinical use.

Tetracycline-controlled systems are based on the use of a tetracycline-sensitive DNA-binding protein of bacterial origin (tetracycline repressor). Most systems use fusion proteins between the tetracycline repressor and elements that control gene expression: i) tet-on and tet-off systems are based on a modified tetracycline repressor fused to a viral transactivator [1, 3], which upon tetracycline binding attach to (tet-on) or detach from (tet-off) the DNA binding site; ii) the tTRKRAB system is based on a tetracycline repressor fused to a KRAB element, which silences a ~3kB region surrounding the DNA binding site [4]. In contrast, the T-REx system is based on binding of the native bacterial tetracycline repressor to a modified CMV promoter (CMVTetO2) [5].

For use in ES cells and experimental cell therapy, it is particularly relevant to understand the limitations of the different tetracycline-regulatable systems. Both, tet-on and tet-off systems tend to interfere with cellular functions and have toxic effects in some cell lines [6, 7], including mouse ES cells [8]. The tet-on system is often leaky and exhibits rather low expression levels [9]. The tet-off system requires continuous application of doxycycline to maintain repression of the transgene, which would be poorly suited for future clinical applications [10]. With the tTRKRAB system, irreversible silencing has been observed during early embryonic differentiation [4], and the system might therefore not be compatible with induction in differentiated ES cells. The most promising approach is the T-REx system, allowing high transgene expression levels without cellular toxicity [5, 9]. Unfortunately it is limited by the poor activity of the CMV promoter in ES cells [11, 12].

We therefore decided to construct a tetracycline-inducible system inspired by the T-REx system. However, we replaced the CMVTetO2 promoter by a promoter expected to show a better activity in ES cells. We opted for a modified version of the elongation factor 1-α (EF1-α) promoter, which is strongly active in ES cells [11, 12]. We expressed the system components using the recently described 2K7 lentivectors [12]. We show that the approach allows robust, rapid and dose-responsive induction of transgene expression in ES cells. Importantly, the system allows in vivo induction of gene expression in implanted ES cells.

MATERIALS AND METHODS

Reagents

Reagents and their sources were as follows: the pcDNA6/TR, pT-Rex-DEST30, pDONR221, and pDONRP4-P1R vectors (Invitrogen, Carlsbad, CA, http://www.invitrogen. com); the pGL3-Basic (); pCMX-44x encoding human truncated CD4 was kindly provided by Didier Trono (Ecole Polytechnique Fédérale de Lausanne, Life Sciences Department); the murine CGR8 embryonic stem cell line (European Collection of Cell Culture); the stromal bone marrow MS5 cell line was provided by Katsuhiko Itoh (Department of Clinical Medical Biology, Kyoto University, Kyoto, Japan) [13]; cell culture media, fetal bovine serum, serum replacement, penicillin, streptomycin, non-essential amino acids, sodium pyruvate (Gibco, Paisley, Scotland); blasticidin (Invitrogen, Carlsbad, CA, http://www.in vitrogen.com); Monoclonal Mouse Anti-human CD4/RPE antibody, clone MT310 (Dakocytomation Denmark A/S); Luciferase Assay System (Promega Corporation, http://www.promega.com); Gateway® clonase enzymes (Invitrogen, Carlsbad, CA, http://www.invitrogen.com).

Vector Constructions

We previously described the construction of the $2K7_{bsd}$ [12]. To generate the $2K7_{tCD4}$ and the $2K7_{GFP}$ lentivectors, the blasticidin resistance coding sequence and the bacterial EM7 promoter were replaced by the truncated CD4 (tCD4) or the eGFP coding sequences, respectively. To generate pENTREF1-αTetO2, top (CGTCCCTATCAGTGAT AGAGATCTCCCTATCAGTGATAGAGA) and bottom (CGTCTCTATCACTGATAGGG AGATCTCTATCACTGATAGGGA) oligonucleo-tides encoding the TetO2 element [5] were annealed and ligated to Acr*I* in pENTREF1-α [12]. To generate entry vectors, the different promoters and genes of interest were cloned into pDONRP4-P1R and pDONR221, respectively, using the Gateway® BP clonase enzyme mix. The resulting entry vectors were then recombined into $2K7_{GFP}$, $2K7_{bsd}$, or $2K7_{tCD4}$ lentivectors using the Gateway® LR plus clonase enzyme mix.

Cell Cultures

CGR8 ES cells were cultured on gelatin-coated dishes in BHK-21 medium supplemented with 10% fetal calf serum, L-glutamine, non-essential amino acids, sodium pyruvate, penicillin and streptomycin, and leukemia inhibitory factor (LIF). HeLa cells were cultured in DMEM low glucose supplemented with 10% fetal calf serum, penicillin and streptomycin.

ES Cell Differentiation

Neuronal differentiation was carried out as described [13]. Briefly, irradiated MS5 cells (1.25×10^5/well) were seeded in 6 well plates. The next day, CGR8 cells ($15\text{-}20 \times 10^3$ cells/well) were plated on the MS5 layer in complete DMEM medium supplemented with non-essential amino acids, 2-mercaptoethanol, 15% knockout serum. Cells were analyzed 4-6 days after plating on the MS5 layer. After this treatment, >95% of the cells were either nestin or β3 tubulin-positive (data not shown). Embryoid bodies were generated by the hanging drop method as described [14].

Lentivector Production and Transductions

Lentivector particles were produced by transient transfection in 293T cells as previously described [15]. To generate transduced CGR8 ES cell lines, lentivector-containing supernatants were collected after 72 hours, filtered through 0.45-μm pore-sized polyethersulfone membrane and concentrated 150 fold by ultracentrifugation ($50'000g$, for 90 min. at 4°C). The pellet was resuspended in complete cell culture medium and was subsequently added to the target cells. $1\text{-}2 \times 10^7$ lentivector particles were used to transduce CGR8 ES cells (2×10^4 cells/well) in suspension on gelatin-coated 96-well plates. The next day, cells were passaged on gelatin-coated 6-well plates. For CGR8 transduced with the $2K7_{bsd}$, blasticidin was added 72 hours after transduction at 7.5 μg/ml and was maintained for 6 days. To generate transduced HeLa cell lines, 2×10^4 cells/well were transduced with 0.2-3ml of the filtered lentivector-containing supernatant in 6-well plates. Cells were split in 85mm culture dishes 72 hours after transduction.

Quantitative Analysis of Cells Expressing Fluorescent Proteins

eGFP-transduced HeLa cells were analyzed by flow cytometry using a FACSscan (Beckton- Dickinson) or a FACSsorter (Beckton-Dickinson). Cell sorting of CGR8 and HeLa cells was performed on a FACSsorter (Beckton-Dickinson).

Luciferase Assays

Cell lysates were prepared according to the manufacturer's instructions. Luminescence measure-ments were performed on a Fluostar Optima (BMG Labtech GmbH, Hanns-Martin-Schleyer-Str. 10, D-77656 Offenburg/Germany).

In Vivo Luminescence Measurements

7-12 weeks old C57/Bl6 mice (Charles-River France) were used. All experiments were approved by the local ethic committee and follow the Swiss guidelines for animal experiments. Gas anaesthesia (isofluran 1-2%) was used during all animal manipulation and imaging sessions. D-Luciferin potassium salt (Xenogen Corporation, Hopkinton, MA, USA) was injected intraperitoneally at a dose of 150mg/kg. Images were acquired on an IVIS-200 system (Xenogen Corporation, Hopkinton, MA, USA) 20 minutes post D-Luciferin injection. Regions of interest were manually drawn. Results are presented as number of photons/second/cm^2.

RESULTS

Engineering of the EF1Tet system

We aimed to engineer an inducible system in ES cells with the following properties: i) absence of toxicity; ii) rapid and dose-responsive inducibility; iii) high absolute expression levels; iv) maintenance of inducibility upon ES cell differentiation. We first investigated the T-REx system (see Introduction) in the CGR8 mouse ES cell line. Cells were transduced with a 2K7$_{bsd}$ lentivector expressing eGFP under the control of the CMVTetO2 promoter, and eGFP expression levels were quantified by flow cytometry. This system did not yield interesting results; indeed, even in the absence of TetR, eGFP expression levels remained very low (data not shown). This is not surprising since the CMV promoter is poorly active in mouse ES cells [11, 12]. In contrast, the human EF1-α promoter has a strong activity in ES cells [11, 12]. We therefore decided to construct an EF1-α promoter-based tetracycline inducible system. As previously described, the native tetracycline repressor can efficiently repress expression of transgenes when the high affinity TetO2 sites are placed 10 or 22 base pairs after the last A of the promoter TATA box [5]. We therefore inserted two high affinity tetracycline operators (TetO2 elements) 22 base pairs after the last A of the TATA box (TATATAA) of the human EF1-α promoter (Figure 1A). In the absence of tetracycline, two TetR dimers remain bound to the TetO2 elements and prevent transcription of the transgene. Addition of tetracycline induces a conformational change in TetR and releases it from the TetO2 elements, enabling full activation of the promoter (Figure 1B). Throughout the text, we will refer to this inducible system as EF1Tet.

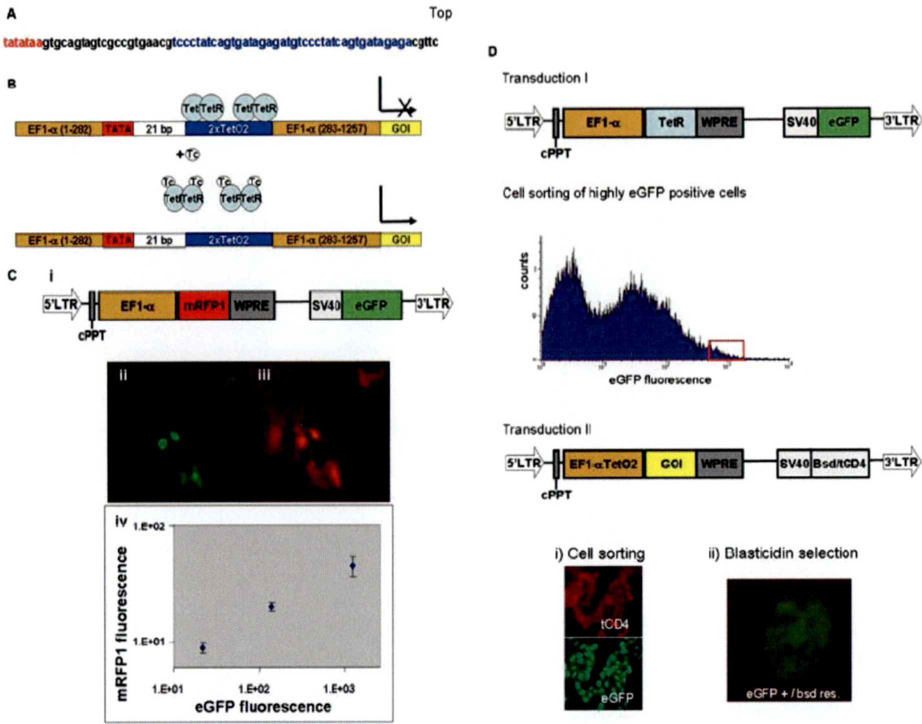

Figure 1. EF1Tet system. A: Two binding sites for tetracycline repressor (TetR) dimers (TetO2 elements, blue) were inserted 22 base pairs downstream of the TATA box (red) of the human EF1-α promoter. B: Illustration of tetracycline-controlled transgene expression. In the absence of tetracycline, two TetR dimers bind to the TetO2 operators, blocking mRNA synthesis. Upon tetracycline addition, TetR dimers are released from the TetO2 operator, enabling gene transcription. C: CGR8 cells were transduced with the 2K7GFP to express mRFP1 under the control of the EF1-α promoter. i: schematic of the 2K7GFP EF1-αmRFP1; ii: eGFP (green); iii: mRFP1 (red); iv: quantification of red fluorescence in three subpopulations of cells expressing eGFP at low, medium and high levels. Data are mean of 3 independent experiments +/- S.D. D: Two lentivectors were required to generate inducible cell lines. A first round of transduction was performed to express TetR controlled by the EF1-α promoter and eGFP controlled by the SV40 promoter. Sorting of highly eGFP positive cells allowed to obtain a cell line with high level TetR expression. A second round of transduction allowed inducible expression of a transgene under the control of the EF1-αTetO2 promoter. Double-transduced cells were selected either by i) cell sorting using a fluorescently-labeled anti-CD4 antibody: HeLa cells double positive for tCD4 (immunostaining) and eGFP; ii) cells after blasticidin (bsd) treatment: fluorescence picture of blasticidin resistant, eGFP positive CGR8 ES cells.

Generation of Inducible HeLa and ES Cell Lines

To generate EF1Tet cell lines, it is necessary to coexpress TetR and the gene of interest controlled by the inducible promoter. In addition, TetR must be expressed at high levels to minimize promoter leakage under non-induced conditions.

We previously described a novel generation of lentivectors enabling rapid generation of ES cell lines harboring virtually 100% of transgene-expressing cells [12]. To be able to optimize TetR expression, we generated two new 2K7 lentivectors by replacing the blasticidin resistance of the 2K7$_{bsd}$ by selection markers that allow cell sorting by flow cytometry. We generated the 2K7$_{GFP}$ and the 2K7$_{tCD4}$, in which eGFP or a truncated human CD4 (tCD4), respectively, are expressed under the control of the ubiquitous SV40 promoter. tCD4 is devoid of its intracellular signaling domain and therefore biologically inactive.

This strategy is based on the assumption that the expression levels of the reporter gene in the vector backbone (e.g. eGFP or tCD4) reflect expression levels of the transgene of interest. To test this assumption, we constructed a vector containing a red fluorescent protein (mRFP1, [16]) under the control of the EF1-α promoter in the 2K7$_{GFP}$ backbone (Figure 1C i). Figure 1C ii-iii shows fluorescence microscopy of CGR8 ES cells transduced with this vector. We then performed quantitative analysis of the correlation between eGFP and mRFP1 expression by flow cytometry analysis (Figure 1C iv). These data demonstrate that sorting of highly eGFP-positive cells allows to select for a subpopulation that expresses high levels of a transgene of interest.

Figure 1D illustrates the generation of inducible EF1Tet cell lines. The 2K7$_{GFP}$ mediates the expression of TetR driven by the human EF1-α promoter in HELA and mouse CGR8 ES cells. After transduction, strongly eGFP positive cells were sorted to obtain a cell line expressing high levels of TetR. Cells were then submitted to a second round of transduction mediated either by the 2K7$_{tCD4}$ or the 2K7$_{bsd}$ to express firefly luciferase under the control of the EF1-αTetO2 promoter (Figure 1D). Finally, cells were either submitted to blasticidin selection or sorted by flow cytometry using a fluorescently labeled anti-human CD4 antibody.

Dose-Responsiveness and Absolute Induction Levels

We first investigated the different EF1Tet HeLa cell lines for their tetracycline-inducible transgene expression. In a first set of experiments, firefly luciferase was used as an inducible reporter transgene because of its very high sensitivity, broad range of detection, and absence of background signal. tCD4 was used as a selection marker for the construct carrying the inducible promoter driving firefly luciferase expression. Cells were cultured for 24 hours in the presence of different concentrations of tetracycline. After 24 hours, cells were lysed and assayed for luciferase activity. Figure 2A shows luminescence detection traces of HeLa cells cultured for 24 hours with 1μg/ml of tetracycline (red trace) or without tetracycline (blue trace). Figure 2B shows dose-responsiveness of luciferase expression level as a function of tetracycline concentrations. Maximal induction (17 fold) was obtained with 1μg/ml of tetracycline. Doxycycline treatment gave similar results (data not shown).

We next investigated the absolute expression levels generated by the inducible EF1-αTetO2 promoter. For this purpose, we generated different HeLa cell lines with inducible or ubiquitous eGFP expression. HeLa cells were first transduced with the 2K7$_{bsd}$EFLTetR. Six cell lines were then generated by subsequent transduction with various amounts of the two lentivectors 2K7$_{bsd}$EF1-αTetO2GFP and 2K7$_{bsd}$EF1-αGFP. Inducible cells were cultured in the presence or absence of tetracycline for 48 hours, and eGFP expression levels were quantified by flow cytometry (Figure 2C). Expression levels driven by the induced EF1Tet system reached 27% of those achieved with the native EF1-α promoter.

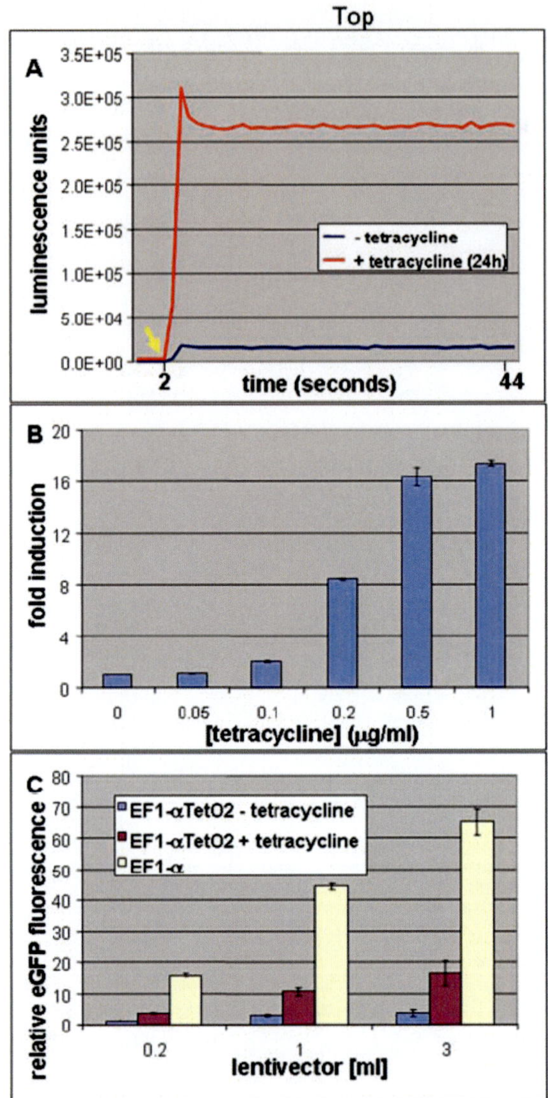

Figure 2. The EF1Tet system allows dose-responsive expression of transgenes. A: Luminescence measurements in HeLa cells. 5x105 HeLa cells were cultured with or without tetracycline (1µg/ml) for 24 hours, subsequently lysed and assayed for luciferase activity. Luciferase substrate was injected 2 seconds after initiation of luminescence measurements (yellow arrow). B: Dose-response of luciferase expression in HeLa cells. After a 24 hours exposure to different doses of tetracycline, cells were assayed for luciferase activity. Data are mean of 2 experiments done in triplicate +/- range. C: HeLa cells constitutively expressing TetR were transduced with different amounts of lentivectors expressing eGFP under the control of either the EF1-α or the EF1-αTetO2 promoter. Cells were cultured with or without tetracycline at 1µg/ml for 48 hours, and were subsequently analyzed for eGFP fluorescence by flow cytometry. Data are mean of 3 independent experiments +/- S.D.

Generation of Inducible CGR8 Mouse ES Cells

Inducible ES cell lines were generated as described in Figure 1. First, CGR8 cells were transduced with the 2K7$_{GFP}$EFLTetR. Three subpopulations expressing eGFP at different levels were sorted by flow cytometry in order to obtain cells with different TetR expression levels. Lines with the highest expression levels showed the best inducibility (data not shown), and were used for further experiments. In one line, the relative induction of luciferase activity after 24h of tetracycline exposure was 11 fold in undifferentiated cells (Figure 3A). For a potential use in cell therapy, inducibility in differentiated ES cells is an important feature. We therefore studied induction of luciferase in neurally differentiated EF1Tet CGR8 cells and found induction levels that were close to values seen with undifferentiated cells (8 fold, Figure 3A).

Another important issue for use of an inducible system in implanted ES cells is the question of the reversibility of transgene expression. To address this question, we used another EF1Tet CGR8 line. This line had a lower relative increase, but high absolute luciferase expression levels. Figure 3B shows that luciferase expression returned to basal levels within 24 hours after removal of tetracycline. Thus, EF1Tet ES cell lines show rapid induction and reversibility, as well as a persistence of inducibility in differentiated cells.

Further analysis of these lines yielded the following results. The different TetR-expressing ES cell lines (n=3) could be propagated and maintained in an undifferentiated state similarly to control cells. The multilineage differentiation potential was confirmed by an unchanged number of i) beating cardiomyocyte clusters during embryoid body differentiation, and ii) β3 tubulin-positive cells during neuronal differentiation on MS5 feeder cells (data not shown).

In Vivo Induction of Luciferase Expression

We next investigated induction of transgene expression in vivo with luciferase as reporter gene. 2x10^6 EF1Tet CGR8 cells were injected subcutaneously in the left ear of C57/Bl6 mice, while 2x10^6 CGR8 expressing TetR alone were injected in the right ear of the same mice. Animals received either doxycycline or PBS intraperitoneally, immediately after cell injection and again after 20 hours. In vivo luminescence measurements were performed after 24 hours. Luminescence in the ears injected with the cells expressing TetR alone was almost undetectable (mostly in the range of 10^2 photons/sec/pixel^2). Low level luminescence emission was detected in ears injected with EF1Tet cells after PBS treatment (Fig 4A and B). In contrast, high luminescence levels were detected in ears injected with EF1Tet cells after doxycycline treatment (Figure 4C and D). Median luminescence per square pixel was 25 fold higher in doxycycline-treated (N=5) compared to PBS-treated mice (N=4) (p<0.05). We therefore conclude that the EF1Tet system can be used to induce robust and rapid transgene expression in implanted ES cells.

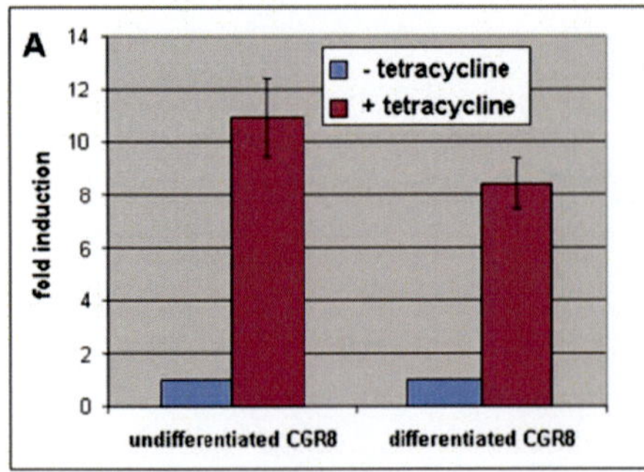
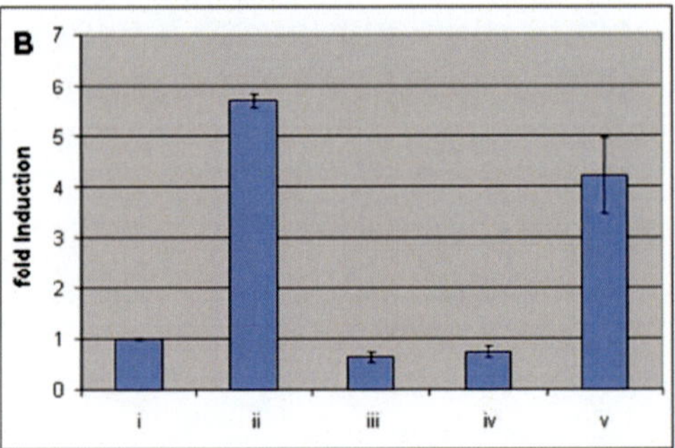

Figure 3. Induction and reversion of transgene expression in ES cells. CGR8 cells expressing high levels of TetR were transduced with a construct allowing inducible firefly luciferase expression.
A: Undifferentiated or neurally differentiated ES cells were assayed for luciferase activity after culture with or without tetracycline. Data are mean of 3 experiments done in duplicates +/- S.D.
B: Reversibility of luciferase expression induction. CGR8 cells with high inducible level of luciferase expression were assayed for luciferase activity in different conditions: i) 24 hours without tetracycline; ii) 24 hours with tetracycline; iii) 48 hours without tetracycline; iv) 24 hours with tetracycline followed by 24 hours without tetracycline; v) 48 hours with tetracycline. Data are mean of 2 experiments done in triplicate +/- range.

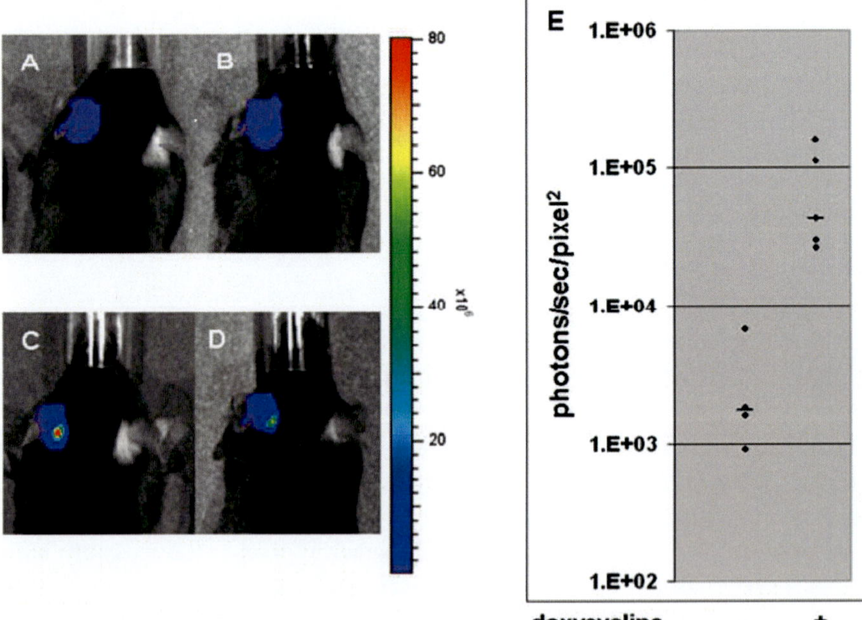

Figure 4. In vivo induction of luciferase expression. 2x106 CGR8 cells inducible for luciferase were injected in the left ear of C57/Bl6 mice. 2x106 control cells expressing TetR alone were injected in the right ear. Five mice received 4mg of doxycycline intraperitoneally immediately after cell injection and again 20 hours later. Four control mice were injected with PBS according to the same time schedule. 24 hours after cell injection, D-luciferin was injected intraperitoneally and in vivo luminescence measurements were performed after 20 minutes. A, B: two PBS-treated mice. C, D: two doxycycline-treated mice. E: Synopsis of luminescence emission in 4 PBS-treated and 5 doxycycline-treated animals. P<0.05. Color bar: min = 4.33x105; max= 8x107.

DISCUSSION

In this article we describe an inducible system based on a modified EF1-α promoter and the native tetracycline repressor. We used modified versions of the recently described 2K7 lentivectors [12] to generate stable inducible cell lines. The characteristics of the system are: i) rapid on/off switch; ii) fine tuning of transgene expression levels; iii) functionality in both undifferentiated and differentiated mouse ES cells; iv) robust and rapid in vivo induction of implanted ES cells.

The establishment of pharmacologically - regulatable systems in embryonic stem cells is a considerable challenge. ES cells are very sensitive to toxic effects of expressed transgenes [8]. Expression of transgenes in ES cells is technically demanding, and the regulatable systems should anticipate ES cell-specific requirements, such as transgene expression after cell differentiation and after implantation. Here we will briefly discuss limiting factors in previously described systems and in the present work.

Cellular toxicity: Systems based on fusion proteins between the tetracycline repressor and viral transactivators have been shown to be toxic in various cell types, including ES cells [3, 6]. We never observed any signs of toxicity, even in cell lines expressing high levels of tetracycline repressor. Furthermore, TetR-expressing ES cells retained the same proliferative and differentiation capacities as untransduced cells.

Ease and stability of transgene expression: To generate inducible ES cell lines, it is particularly important i) to minimize transgene silencing, and ii) to control TetR expression level to optimize inducibility. 2K7 lentivectors are well suited for this purpose, allowing rapid generation of polyclonal stable cell lines with maintenance of transgene expression during differentiation [12]. In addition, the 2K7$_{GFP}$ allows to select for a subpopulation with high TetR expression levels to optimize inducibility of the cell lines.

Inducibility in differentiated cells: Maintenance of transgene inducibility during ES cell differentiation is of critical importance for developmental studies and future cell therapy. Inducible systems may lose transgene induction during cell differentiation. Indeed, the tTRKRAB fusion protein-based system [4] allows switch on of transgene expression at later developmental stages only if doxycycline is added during the first days of embryonic development [4]. This is a major limitation since the necessity for early induction of a given transgene may have dramatic effects on developmental events. A recently published paper [17] shows the establishment of the tTRKRAB-based system in human embryonic stem cells. Efficient induction was achieved in undifferentiated cells; however, inducibility after cell differentiation was not investigated. In the EF1Tet system, transgene expression can be rapidly switched on in already differentiated neurons. This is also a major advantage for future applications in cell therapy, where the possibility to induce transgenes in fully differentiated cells is a desirable feature.

Inducibility after cell implantation: In vivo induction of transgene expression is of major interest for experimental cell therapy and potential clinical use. Tetracycline-regulatable systems are particularly well-suited for this purpose because of the safety of the inducer compound. The previously described tet-off system has also shown the ability to regulate transgene expression in implanted ES cells [18]. However, this system has a low potential for a later clinical use, because of the need to continuously administrate doxycycline to repress transgene expression. In addition, the system responds slowly (i.e. a time range of several days) to changes in tetracycline concentration [18]. The system described here allows in vivo induction of transgene expression within 24 hours, thus necessitating only a short time period of drug treatment.

One of the ultimate goals of inducible transgene expression is its use for gene and cell therapy. A recently published review [10] mentions a series of criteria that pharmacologically-regulated systems should fulfill to be suitable for clinical use. Table 1 summarizes to which extent these criteria are met by different tetracycline-regulated systems [1, 3-5, 9, 18, 19]. Given many similarities, our EF1Tet system is shown in the table together with the T-REx system approach [5]. Note however that because of the low activity of the CMV promoter in ES cells [11, 12], the T-REx system would not be suitable for most ES cell applications. We show here that the EF1Tet system, i.e. EF1-αTetO2 promoter used in conjunction with the native bacterial TetR, satisfies most of these criteria. Additional advantages include rapid on/off switch of transgene expression, maintenance of inducibility after ES cell differentiation, and in vivo induction of transgenes. However, the issue of TetR immunogenicity should be carefully addressed to allow the use of this system in humans. The impact of promoter leakage should also be investigated for each transgene of interest.

Table 1.

	Tet-on	Tet-off	tTRKRAB	EF1Tet/T-REx
On switch	YES	NO	YES	YES
Absence of interference with endogenous pathways [7]	NO	NO	Not reported	Not reported
Oral bioavailability of the inducer	YES	YES	YES	YES
Safety of the inducer compound	YES	YES	YES	YES
Reversibility of induction [1, 3-5, 17]	YES	SLOW	YES	YES
Low basal activity [4, 9]	NO	YES	YES	NO
High inducibility [4, 5, 9, 17]	NO	YES	YES	YES
Dose dependence [4, 5, 9, 17]	YES	YES	NO	YES
Low immunogenicity [18]	NO	Not reported	Not reported	Not reported

In conclusion, the system presented here is well suited for transgene induction in undifferentiated and differentiated ES cells. The system is functional in vitro and in vivo. Being a switch on inducible system, it is particularly well suited for in vivo applications.

ACKNOWLEDGMENTS

We would like to thank Dominique Wohlwendt and Susanne Bissat for help with flow cytometry procedures, Christine Deffert for help with subcutaneous injections in mice, and Prof. Ueli Schibler for helpful discussion and advice. This work was supported by grants from the Swiss National Science Foundation (NFP46, 404640-101109) and the Clayton Foundation. D.S. is a recipient of an M.D./Ph.D. scholarship from the Swiss National Science Foundation.

REFERENCES

[1]   Gossen M, Bujard H. Tight control of gene expression in mammalian cells by tetracycline-responsive promoters. *Proc. Natl. Acad. Sci. USA* 1992; 89: 5547-51.

[2]   Shockett P, Difilippantonio M, Hellman N et al. A modified tetracycline-regulated system provides autoregulatory, inducible gene expression in cultured cells and transgenic mice. *Proc. Natl. Acad. Sci. USA* 1995; 92: 6522-6.

[3]   Gossen M, Freundlieb S, Bender G et al. Transcriptional activation by tetracyclines in mammalian cells. *Science* 1995; 268: 1766-9.

[4]   Szulc J, Wiznerowicz M, Sauvain MO et al. A versatile tool for conditional gene expression and knockdown. *Nat. Methods* 2006; 3: 109-16.

[5]   Yao F, Svensjo T, Winkler T et al. Tetracycline repressor, tetR, rather than the tetR-mammalian cell transcription factor fusion derivatives, regulates inducible gene expression in mammalian cells. *Hum. Gene Ther.* 1998; 9: 1939-50.

[6]   Gallia GL, Khalili K. Evaluation of an autoregulatory tetracycline regulated system. Oncogene 1998; 16: 1879-84

[7]   Baron U, Gossen MBujard H. Tetracycline-controlled transcription in eukaryotes: novel transactivators with graded transactivation potential. *Nucleic Acids Res.* 1997; 25: 2723-9.

[8]   Bryja V, Pachernik J, Kubala L et al. The reverse tetracycline-controlled transactivator rtTA2s-S2 is toxic in mouse embryonic stem cells. *Reprod. Nutr. Dev.* 2003; 43: 477-86.

[9]   Xu ZL, Mizuguchi H, Mayumi T et al. Regulated gene expression from adenovirus vectors: a systematic comparison of various inducible systems. *Gene* 2003; 309: 145-51.

[10]   Toniatti C, Bujard H, Cortese R et al. Gene therapy progress and prospects: transcription regulatory systems. *Gene Ther.* 2004; 11: 649-57.

[11]   Zeng X, Chen J, Sanchez JF et al. Stable expression of hrGFP by mouse embryonic stem cells: promoter activity in the undifferentiated state and during dopaminergic neural differentiation. *Stem Cells* 2003; 21: 647-53.

[12]   Suter DM, Cartier L, Bettiol E et al. Rapid generation of stable transgenic embryonic stem cell lines using modular lentivectors. *Stem Cells* 2006; 24: 615-23.

[13]   Itoh K, Tezuka H, Sakoda H et al. Reproducible establishment of hemopoietic supportive stromal cell lines from murine bone marrow. *Exp. Hematol.* 1989; 17: 145-53.

[14]   Li J, Puceat M, Perez-Terzic C et al. Calreticulin reveals a critical Ca(2+) checkpoint in cardiac myofibrillogenesis. *J. Cell Biol.* 2002; 158: 103-13.

[15]   Dull T, Zufferey R, Kelly M et al. A third-generation lentivirus vector with a conditional packaging system. *J. Virol.* 1998; 72: 8463-71.

[16]   Campbell RE, Tour O, Palmer AE et al. A monomeric red fluorescent protein. *Proc. Natl. Acad. Sci. USA* 2002; 99: 7877-82.

[17]   Zhou BY, Ye Z, Chen G et al. Inducible and Reversible Transgene Expression in Human Stem Cells after Efficient and Stable Gene Transfer. *Stem Cells* 2006.

[18]   Adachi K, Kawase E, Yasuchika K et al. Establishment of the gene-inducible system in primate embryonic stem cell lines. *Stem Cells* 2006; 24: 2566-72.

[19] Latta-Mahieu M, Rolland M, Caillet C et al. Gene transfer of a chimeric trans-activator is immunogenic and results in short-lived transgene expression. *Hum. Gene Ther.* 2002; 13: 1611-20.

In: Stem Cell Research Progress
Editor: Prasad S. Koka, pp. 95-116

ISBN: 978-1-60456-308-5
© 2008 Nova Science Publishers, Inc.

Chapter 7

BONE MARROW AND ADIPOSE TISSUE-DERIVED MESENCHYMAL STEM CELLS: HOW CLOSE ARE THEY?

Leandra S. Baptista[1,2], Carolina G. S. Pedrosa[2],
Karina R. Silva[2], Ivone B. Otazú[1,2], Christina M. Takiya[1],
Hélio S. Dutra[1,2], César Cláudio-da-Silva[3],
*Radovan Borojevic[1,2] and M. Isabel D. Rossi[1,2]**

[1] Histology and Embryology Department, Biomedical Sciences Institute,
Federal University of Rio de Janeiro, Rio de Janeiro, RJ, Brazil
[2] APABCAM, Clementino Fraga Filho University Hospital,
Federal Univerity of Rio de Janeiro, Rio de Janeiro, RJ, Brazil
[3] Medical Clinics Department, Clementino Fraga Filho University Hospital,
Federal University of Rio de Janeiro, Rio de Janeiro, RJ, Brazil

ABSTRACT

Mesenchymal stem cells (MSC) can be isolated from virtually all adult and fetal tissues and expanded in vitro. They share their origin in a perivascular niche suggesting a relationship with pericytes. Adult subcutaneous tissue is an abundant source of MSC, which share with bone marrow mesenchymal stem cell (BM-MSC) not only the capacity to differentiate into different mesenchymal lineages, but also an immunomodulatory effect. In view of the capacity to differentiate into different mesodermal tissues, the potential application of adipose tissue-derived MSC (ATMS) in regenerative medicine as an alternative source to BM-MSC has raised great interest. In this chapter, we will discuss current data that suggest differences in phenotype, gene expression profile, osteo-

---

* Corresponding author: Maria Isabel Doria Rossi. Depto. Histologia e Embriologia, Instituto de Ciências Biomédicas, Universidade Federal do Rio de Janeiro (UFRJ). Sá Ferreira, 227 / 1005, Copacabana, Rio de Janeiro, 22071-100, RJ, Brazil. Tel: (55) (21) 2562-2468. Fax: (55) (21) 2562-2467. E-mail: idrossi@hucff.ufrj. br

chrondogenic potential, and hematopoietic supportive property between AT-MSC and BM-MSC.

Keywords: mesenchymal stem cell, adipose tissue, bone marrow, pericytes, hematopoiesis

INTRODUCTION

Adult bone marrow (BM) is traditionally viewed as a hematopoietic organ, that is, a tissue whose principal function is to provide a microenvironment that supports the proliferation and differentiation of hematopoietic stem cells (HSC) throughout life. The composition of the BM stroma is complex. The nonhematopoietic components comprise advential cells, commonly referred to as "reticular cells", residing on the abluminal side of the endothelium with long processes emanating from the blood vessel wall into the adjacent tissue, vascular smooth muscle and endothelial cells, adipocytes, and osteogenic cells near bone surface [1-2]. Pioneering studies isolated from adult bone marrow stroma adherent, clonogenic, fibroblast-like cells (known as CFU-F, colony forming unit-fibroblast), which could give rise under appropriate experimental approaches to different mesenchymal tissues, including bone, cartilage, and adipose tissue [3-7]. These data raised the hypothesis that bone marrow stroma was itself based on a stem cell hierarchical model.

The observation of considerable variation in proliferation capacity and developmental potential among individual colonies [8-10] supports this idea. According to this model, all BM stromal cell lineages, including the myelosupportive stroma, are derived from a common multipotent, quiescent, self-renewed stem cell referred in contemporary literature as mesenchymal stem cell (MSC) or bone marrow stromal stem cell. Consequently, adult BM must be viewed nowadays as an organ that harbors two distinct stem cell populations that not only co-exist but are also functionally interdependent [2].

Recently, the differentiation capacity of BM-MSC has been widely investigated. Numerous studies suggest that BM-MSC differentiate into cardiomyocytes [11] and contribute to cardiac function recovery [12-13], improve nerve regeneration [14], cartilage and bone repair [15], angiogenesis [16], and hematopoietic engraftment, and have also immunosuppressive property [reviewed in 17]. This astonishing "plasticity" of BM-MSC has not surprisingly generated considerable interest on potential therapeutic application of these cells and have also led to the search of alternative sources of MSC.

MSC-like cells have been isolated from different tissues – e.g. adipose tissue [18], blood [19-20], synovium [21], dermis skeletal muscle [22], dental pulp and periodontal ligament [23], placenta [24], and human scalp tissue [25]. More recently, it was shown that mesenchymal stem cells might reside in virtually all adult tissues [26]. Whether MSC derived from these different sources represent fundamentally similar cell types is not clear.

ADIPOSE TISSUE: AN ALTERNATIVE TO BONE MARROW AS A SOURCE OF MSC?

Adipose tissue-derived MSC (AT-MSC), also called adipose tissue-derived stromal vascular fraction cells (SVF), are of great interest. Adipose tissue is an abundant and accessible source of MSC that can be easily isolated from this tissue and expanded *in vitro*. Their relative frequency *in vivo* is thought to be higher [18] than the frequency of MSC in BM, which ranges from 0.1 to 5 x 10^{-5} cells [27]. We also observed that the relative frequency of adherent cells in adipose tissue and lipoaspirates after enzymatic digestion reaches 10^{5} cells / g of tissue. The procedure to obtain BM-MSC cell population – e.g. great volume of BM aspirates from iliac crest are required – might cause discomfort to the patients. On the contrary, adipose tissue is easily accessible. Finally, *ex vivo* expanded AT-MSC showed extensive capacity to differentiate *in vitro* and/or *in vivo* into bone, cartilage, muscle [18, 28-31], and endothelium [32]. It is also striking that AT-MSC seem to share with BM-MSC an immunosuppressive property [33-35].

The reciprocal relationship in the osteogenic and adipogenic differentiation pathways of BM-MSC [36-37] and the evidence in a pathological condition named osseous heteroplasia that subcutaneous adipose cells can convert to osteoblasts-like cells leading to ectopic bone formation in subcutaneous fat [38-39] suggest a close relationship between BM-MSC and AT-MSC. However, few comparative functional studies have been done.

Microarrays analysis showed that the gene expression profile of BM-MSC and AT-MSC is very similar, although minor differences were observed [40]. It should be said that the profile was also modified by the culture conditions, which could cause subtle differences observed in differentiation of the cells expanded *in vitro*. It was also shown that the choice of serum modifies proliferation, differentiation, and gene transcription profile of human BM-MSC [41]. Actually, culture conditions can affect cell growth and differentiation in many ways. Usually, MSC are isolated by their plastic adherence property and are expanded onto plastic dishes. This is a caveat, since it disrupts the complex tridimensional (3-D) interactions between cells and extracellular matrix (ECM), modifying cytoskeleton organization and cell shape. Since cell shape markedly influence cell proliferation, differentiation, and gene expression [42-46], it is not surprising that minor changes in culture conditions would affect the gene expression profile of MSC. A very interesting finding that illustrates how 3-D ECM affects terminal differentiation of fat cells was recently reported [47]. The authors noticed that adipocytes maturation *in vivo*, but not in two-dimensional (2-D) cultures, requires the expression of the metalloproteinase MT1-MPP. In the absence of this protease, white adipose tissue development was aborted but null progenitors were able to give rise to fully differentiated adipocytes in 2-D cultures. In conclusion, although data obtained from culturing cells in 2-D *in vitro* system should be viewed with caution, they are still valuable.

Dicker and co-workers [48] claimed that adipocytes obtained by *in vitro* differentiation of AT-MSC are fully functional and virtually indistinguishable from fat cells derived from BM-MSC. However, it was recently shown that lipid metabolism related genes are upregulated higher in AT-MSC than BM-MSC during adipogenic differentiation, suggesting that AT-MSC are superior to BM-MSC in adipogenesis [49]. Others subtle differences in the osteo-chondrogenic potential of these cells were recently reported [29-30, 49-50]. The two MSC, AT-MSC and BM-MSC, undergo chondrogenic differentiation when cultivated in 3D-high density cultures in a chondrogenic medium. However, differences in the expression of extracellular matrix genes during differentiation suggested that BM-MSC are superior to AT-MSC in chondrogenesis [49], which is in agreement with previous observation that BM-MSC but not AT-MSC reached gene expression profile of fully differentiated chondroblasts [30]. Transplantation assays where collagen-I sponges [29] or poly-L-lactic acid [51] were seeded with AT-MSC or BM-MSC showed that BM-MSC efficiently formed new bone tissues, while this was not achieved by AT-MSC unless osteogenic differentiation had been induced either by BMP-2 transfection [29] or by cultivation in osteogenic medium [51]. Collagen-I and poly-L-lactic acid are not osteoinductive biomaterials and the use of other scaffolds that provide a more suitable environment, as hydroaxiapatite tricalcium phosphate (HA-TCP), could have an osteogenic inductive role favoring the development of osteoid tissue by AT-MSC as described [52]. The recent observation [49] that during *in vitro* osteogenic differentiation, BM-MSC expressed higher levels of bone extracellular matrix genes as compared to AT-MSC is in agreement with the *in vivo* data.

Taken together these results suggest a close relationship between BM-MSC and AT-MSC and although minor differences concerning the adipogenic and osteo-chodrogenic potential of these cells have been detected, it is not clear whether these differences are related to the environment from where the cells are derived, to subtle *in vitro* differentiation that might lead to variation in the relative frequency of committed progenitors, or to their different origin.

PHENOTYPIC CHARACTERISTICS OF BONE MARROW AND ADIPOSE TISSUE MESENCHYMAL STEM CELLS

The phenotypic characterization of BM-MSC and AT-MSC is mostly based on analysis of *ex vivo* expanded cells and this can be one of the major causes of the differences described by different groups. None of the markers is specifically expressed by mesenchymal cells that have been isolated based on their capacity to adhere on plastic surfaces. Moreover, different methods of isolation and expansion have been described and different approaches to characterize these cells have been used. Thus, it is extremely difficult to compare the outcomes of different studies and a temptative definition of MSC based on minimal criteria was recently proposed by The International Society for Cellular Therapy (ISCT) [53]. According to this statement, MSC must: (i) be plastic-adherent in standard cultures; (ii) express CD105, CD73 and CD90, and lack expression of CD45, CD34, CD14 or CD11b, CD79α or CD19 and HLA-DR; and (iii) differentiate into osteoblasts, adipocytes and chondroblasts *in vitro*.

The proposed phenotype characteristics were mainly based on BM-MSC description. In fact, human BM-MSC have been characterized as adherent cells that lack hematopoietic markers (CD45, CD14, HLA-DR, CD11b, and CD34), and express CD29, CD44, CD106, CD105, CD73, CD90 [27,53], and STRO-1, since selection of STRO-1[+] cells from fresh BM aspirates resulted in a 10- to 20- fold enrichment in CFU-F compared to unseparated BM [54]. Although AT-MSC show a similar phenotype, some distinguishing characteristics have been reported.

AT-MSC also express CD13, CD29, CD44, CD90, CD105, and STRO-1 and lack the expression of the hematopoietic cell markers CD45, CD14, and HLA-DR [18, 40,55], but differences regarding the expression of adhesion molecules have been reported. Zuck and colleagues [18], as wells as De Ugarte and colleagues [55], observed that AT-MSC were positive for CD49d while BM-MSC were not. On the other side, no expression of CD49d in both cells was reported [40]. We, and others [25, 56], observed the expression of CD49d in both cells (Figure 1M and unpublished data).

The expression of CD34 is of particular interest because it is well established in the literature that BM-MSC do not express this marker and, as pointed out above, the absence of CD34 expression was even proposed as one of the phenotypic characteristics of MSC by the ISCT [53]. However, there is no consensus in the literature in relation to the expression of CD34 by AT-MSC [18, 32, 40, 57-58]. We observed that cells isolated from enzimatically digested lipoaspirates or adipose tissue are heterogeneous and contains hematopoietic (CD45$^+$) and nonhematopoietic (CD45$^-$) cells (Figure 1A-B). The percentage of CD45$^-$ cells was variable (40.9% ± 21.5) and virtually all of them were CD34^{high}, but just a fraction (ranging from 9.9% to 31.9%) was CD105$^+$. The CD105$^+$ cells were also CD34$^+$ (Figure 1C). After 72 hours of culture, the plastic-adherent CD45$^-$ cell population (Figure 1F) was smaller and less granular (Figure 1D) than the cells harvest when they were confluent (Figure 1E). Besides, the expression of CD34 and CD105 was still very similar to the freshly isolated cells, that is, most of them were CD34^{high} and a small fraction of these cells were also CD105$^+$ (Figure 1G). However, down-modulation of CD34 (the MFI – mean fluorescence intensity – dropped from 536.3 to 75.1) that was accompanied by a subtle upregulation of CD105 (the MFI changed from 21.8 to 49.2) was observed as cells reached confluence (Figure 1H-I). Furthermore, the expression of CD34 was lost during *in vitro* expansion (Figure 1K), as observed previously [56-59], and the cells became homogenously positive for CD105 (Figure 1L). Cells expanded up to passage four were positive for CD105, CD13, and CD49d, did not express hematopoietic (CD45, CD14, and HLA-DR) and vascular (CD31) antigens (data not shown), have lost CD34 expression, and showed the potential to differentiate towards the osteogenic, chondrogenic and adipogenic lineages (Figure 2). Although we can not rule out the possibility of expansion of the CD34$^-$ cells, the fact that the starting CD105$^+$ cell population was fully CD34$^+$ favors the hypothesis of down-modulation of this molecule during *in vitro* expansion and could explain why there is no agreement in relation to this marker. Differences in size and surface antigens expression were also observed in cultured BM-MSC [60]. Furthermore, it was reported that most of the CD34$^+$CD105$^-$CD31$^-$ cells isolated from human internal thoracic artery wall, that are able to differentiate towards endothelial and smooth muscle cells, acquired CD105 expression after *in vitro* expansion [61].

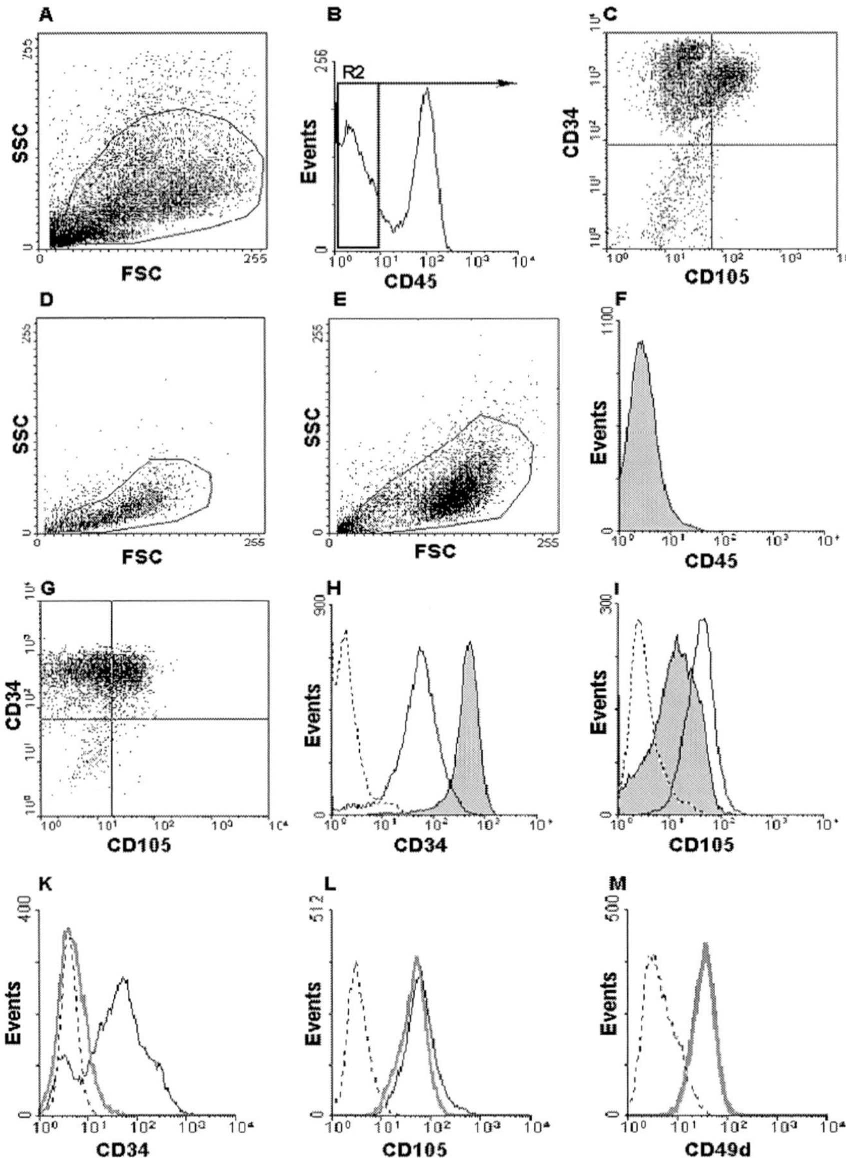

Figure 1. Immunophenotype profile of fresh and cultured adipose tissue derived mesenchymal stem cells. (A) FSC/SSC dot plot of fresh isolated cells. (B) Histogram showing the expression of CD45 by fresh cells. R2 represents CD45- cells. (C) Expression of CD34 and CD105 by fresh CD45- cells (R2). (D-E) FSC/SSC dot plots of adherent cells harvested 72 h after plating or at confluence, respectively. (F) Expression of CD45 by adherent cells. (G) Expression of CD34 and CD105 by adherent cells harvest 72h after plating. (H-I) Histograms showing the expression of CD34 and CD105, respectively, by adherent cells harvest 72h after plating (solid histograms) or at confluence (solid black line). FACS acquisition was done at the same time, so the fluorescence intensity could be compared. (K-M) Histograms showing the expression of CD34, CD105, and CD49d, respectively, by adherent cells harvested at confluence (solid black line) and at passage four (gray solid line). Dashed lines represent negative controls (isotype controls).

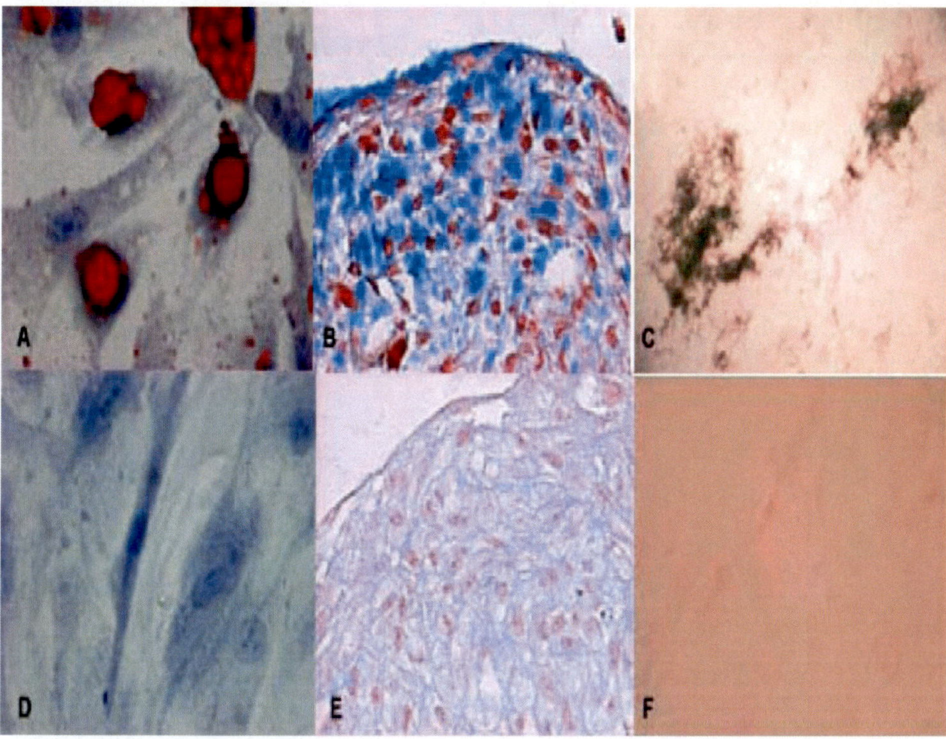

Figure 2. Adipose tissue derived mesenchymal stem cells differentiate to mesenchymal lineages. Cultured adherent cells were harvested after four passages in vitro and maintained with standard adipogenic, chondrogenic or osteogenic medium for up to four weeks. (A) Adipogenic differentiation. Oil Red. 400X. (B) Chondrogenic differentiation. Alcian blue. 200X. (C) Osteogenic differentiation. Von Kossa. 200X. (D-F) Cells were maintained in standard culture medium as negative controls. (D) Oil Red. 400X. (E) Alcian blue. 200X. (F) Von Kossa. 200X.

ORIGIN OF BONE MARROW MESENCHYMAL STEM CELLS: THE RELATIONSHIP WITH VASCULAR MURAL CELLS

A significant obstacle to the study of BM-MSC is the lack of precise knowledge about their *in vivo* location and their ontogeny. While the ontogeny of HSC has been clearly demonstrated [for a review see 62], the origin of marrow stromal cells is still a matter of debate.

BM stromal tissue is formed after chondrolysis of the cartilage rudiment that is followed by vascular, perichondrial, osteoblast and osteoclast precursor cells invasion of the developing bone marrow cavity. A vascular bed is then established between the ossifying trabeculae before any sign of hematopoiesis is observed [63]. Since vascular invasion of the marrow cavity is preceded by calcification of the bone collar, a role of perichondrium in endochondral ossification and vascular invasion was suggested [64]. The presence of cells in adult perichondrium that exhibit the characteristics of MSC, which also support osteoclastogenesis, hematopoiesis, and angiogenesis, sustains the hypothesis that perichondrial cells participate in vascular invasion by recruiting osteoclasts and vessels, and also suggests that perichondrium might serve as a stem cell reservoir playing an important role in the early development of bone and bone marrow [64].

However, as pointed out above, the most striking feature of BM development is the formation of logettes in the marrow cavity that are characterized by a central arteriole surrounded by a loose connective tissue.

These logettes are limited from the sinus by an endothelium layer and at the abluminal side of the endothelium by α-SMA$^+$ cells [63]. Strikingly, bone marrow STRO-1^{bright}CD106$^+$ cells, which are enriched in CFU-F, are also α-SMA$^+$ [10]. *In vivo*, the distribution of α-SMA$^+$ cells in the BM microenvironment is limited to the media of arteries, the abluminal face of sinusoids, and only occasionally to the subendosteal region [2, 27]. The idea that BM-MSCs are derived from vessel-associated progenitors, which might originate from a common ancestor, named mesoangioblast, that would leave the vessel and adopt the fate of BM tissue was then proposed [2, 65-66]. This is in agreement with the observation that STRO-1$^+$ cells are associated with the microvasculature of human BM [23]. Furthermore, cultured human BM stromal cells were shown to express cytoskeletal extracellular matrix proteins specific for vascular smooth muscle cells (vSMC) in a time-dependent manner [reviewed in 9]. The blood vessel-associated progenitor cells that originate the BM stroma might also be a source of perichondrium and periosteoum during osteogenesis [67].

ORIGIN OF ADIPOSE TISSUE MESENCHYMAL STEM CELLS

While the origin and localization of BM-MSC is still a matter of debate, information about AT-MSC is almost absent. The origins of adipose cells and adipose tissue are still poorly understood, and the molecular events leading to the commitment of the embryonic stem cell precursors to the adipocyte lineage remain to be characterized. In most species, white adipose tissue formation begins before birth, as assessed by morphological studies performed on human, pig, mouse, and rat embryos, and its expansion takes place rapidly after birth as a result of increased fat cell size and number [reviewed in 68].

Although the developmental origin of fat cells is not known, several studies on multipotent clonal cell lines have suggested that the adipocyte lineage derives from an embryonic stem cell precursor with the capacity to differentiate into the mesodermal cell types of adipocytes, chondrocytes, osteoblasts, and myocytes [68]. Nowadays, it is currently accepted that adipose lineage arises from multipotent stem cell population of mesodermal origin that resides in the vascular stroma of adipose tissue [reviewed in 69]. During adult life, these precursor cells undergo a multi-step differentiation process comprehending an initial commitment step, in which the cells do not express specific markers, and a subsequent program that results in the acquisition of the adipocyte phenotype [69]. The initial step is likely to be induced by factors secreted by adipocytes undergoing hypertrophy, since for successful *de novo* fat formation, adipose precursor cells need to be transferred together with mature adipocytes and the microenvironment of their regenerative proliferation needs to be maintained [70]. However, since no specific markers for MSC exist, as discussed above, and immunofluorescence studies of adipose tissue are extremely hard to be done, an *in situ* demonstration of these precursor cells in adults has never been shown. The vascular localization of AT-MSC, as proposed, suggests a relationship with pericytes, and the recognition that pericytes may act as a source of immature adipocytes [71-72] together with the observation that neovascularization of basement membrane proteins (matrigel) precedes fat formation [70] strengthens this hypothesis.

PERICYTES AND VASCULAR SMOOTH MUSCLE CELLS

Pericytes and vSMC reside at the interface between the endothelium and the surrounding tissue. The arteries and veins walls are formed by single or multiple layers of vSMC, whereas small capillaries are partially covered by solitary cells referred to as pericytes [reviewed in 73-74]. Besides their role in vascular stability, it has been shown that pericytes isolated from different tissues are actually multipotent progenitor cells, being able to give rise to adipocytes, osteoblasts, and chondroblasts [75-77]. Curiously, like MSC, mural cells isolated from different vessels show differences in the potential to differentiate towards these mesenchymal lineages, which strengthens the hypothesis that MSC and mural vessel cells are related to each other and that fate decision are influenced by local cues. For instance, bovine retinal microvessels-derived pericytes were shown to share some properties with BM stromal cells, including the expression of STRO-1 and the capacity to form bone, cartilage and fibrous tissue *in vivo* [75-77], as well as to differentiate into adipocytes [77]. However, Tintut and colleagues [76] claimed that bovine aortic smooth muscle cells did not differentiate into adipocytes. Accordingly, MSC-like cells isolated from different tissues of adult mice (BM, adipose tissue, muscle, spleen, kidney and glomerulus, lung, liver, brain, pancreas, aorta, and vena cava) although being very similar in their differentiation potential showed differences in the frequency of differentiated cells, as well as in the degree of differentiation [26]. As proposed before, these differences might be related to the local influence of the microenvironment although differences in the ontogeny of these cell populations could not be disregarded.

Pericytes and vSMC may have multiple origins [reviewed in 73-74]. For example, it was shown that pericytes and vSMC of the blood vessels of face and forebrain are derived from neural crest cells that detach from the neural folds and migrate between the ectoderm and the neuroepithelium encountering mesoderm-derived endothelial precursors [78]. Coronary vessel mural cells, on the other hand, are derived from epicardial cells [79]. Differentiation of endothelial cells into pericyte has also been proposed [80], although this process appears to be rare.

Most commonly, pericytes seems to be derived from mesoderm. Cells surrounding the dorsal aorta have been suggested as precursors of the mural cells of trunk vessels in the axial and lateral plate mesenchyme [81-82]. Minasi and colleagues [83] showed that dorsal aorta-derived cells, the mesoangioblasts, incorporate into developing vessels and contribute to mesodermal tissues, including blood, cartilage, bone, smooth, skeletal, and cardiac muscles, which means that pericytes could also be derived from the mesoangioblasts. Recently, a very elegant study where quail somite were transplanted into chick embryos showed that mesoderm somite-derived cells replace the hemangioblasts at the floor of the aorta and provide vSMC of the aorta and trunk arteries. These vSMC precursors colonize the aorta from the medioventral part of the somite, in close association with the sclerotomal compartment [84], but the relationship between these somite-derived cells and the mesoangioblasts stem cells is currently unknown. Curiously, it was thought that osteogenic, chondrogenic, and adipogenic cells may arise from sclerotomal cells whose restricted expression of basic-helix-loop-helix (bHLH) transcription factors (Twist and Scleraxis) may be regulators within the mesenchymal cell lineages [68].

Finally, recent data suggest that in adults, BM-derived cells can migrate to developing vessels and differentiate into pericytes [reviewed in 85]. Although the identity of the cells that give rise to pericytes is not well established and some controversy exists in the literature, the existence of BM-derived pericytes implies that the precursor cells are able to circulate. This observation renews the question as to whether multipotent mesenchymal progenitor-cells are tissue-resident stem cells established during embryonic development or are recruited from the BM or another mesodermal-derived tissue. The available information suggests that circulation of these cells occurs to some extent in adult life and during embryogenesis. In fact, the presence of MSC-like cells has been shown in circulating blood in adults [19-20], cord blood [40, 86-87], and fetal blood from second- and third- trimester [87]. Trafficking of MSC during pregnancy that resulted in fetomaternal microchimerism of BM stromal cells was described [88]. Actually, it was recently proposed that MSC circulate between hematopoietic sites during fetal development, but it is striking that MSC isolated from these sites showed, like MSC isolated from adult tissues, differences in their differentiation potential towards the three mesenchymal lineages [89], suggesting a major role of microenvironment cues in the differentiation capacity of these cells.

The contribution of circulating progenitor cells to the development of adipose tissue during obesity was first proposed by Hong and colleagues [90] that showed that fibroblast-like cells isolated from human peripheral blood were able to differentiate into adipocytes *in vitro* and *in vivo*, forming human adipose tissue following injection into the subcutaneous of immunodeficient mice. A recent elegant and intriguing study [91] demonstrated beyond doubt that BM-derived circulating progenitor cells migrate to adipose tissue and differentiate into adipocytes during obesity. By transplanting whole BM cells from GFP-expressing transgenic mice into wild-type mice and subjecting them to treatment with thiazolidinedione or a high-fat diet, the authors were able to show the presence of GFP$^+$ adipocytes in white and brown adipose tissue. It is interesting that even the controls mice showed few GFP$^+$ adipocytes. The identity of this progenitor cell and its contribution to the tissue-resident MSC population are still a matter of debate. However, the data bring the question to whether MSC isolated from adult adipose tissue are bonafide adipose-derived MSC.

MSC AS A HEMATOPOIETIC SUPPORTIVE STROMA

As pointed out before, a unique characteristic of adult bone marrow stromal cells is their ability to support hematopoiesis, which means that BM-MSC not only differentiate into the several mesenchymal lineages discussed above, but are also able to organize a hematopoietic microenvironment that maintains HSC self-renewal, proliferation, and differentiation. In fact, *in vivo* transplantation of BM-MSC has proved not only the osteogenic potential of these cells, but also their role in developing a hematopoietic tissue, since along the ectopic bone formation, a marrow cavity with hematopoietic cells also develops [2, 23]. These results suggest that BM-MSC may have a role in the organization of BM hematopoietic niches.

Supporting this hypothesis, it was recently shown that ectopic bone formed after transplantation of human BM-MSC into immunocompromised mice subcutaneously with HA-TCP as a carrier contain functional multiple hematopoietic cell lineages, including myeloid, lymphoid, erythroid, and progenitor cells [92]. It is worth to mention that injection of human BM-MSC into the marrow cavity of NOD/SCID mice increased the engraftment of cord blood HSC. Furthermore, human HSC but not their progeny associated specifically with human BM-MSC [93], indicating a particular interaction between the two cell populations.

The presence of MSC in hematopoietic sites during embryogenesis and the isolation of clones derived from different fetal tissues that were shown to differentiate into a myelosupportive stroma [94-95] suggest that this might be an intrinsic characteristic of MSC. Stromal cell clones isolated from AGM subregions were α-SMA$^+$ and showed the capacity to support HSC. However, differences in the hematopoietic supportive capacity of these clones were observed. The urogenital ridges of the AGM region contained highly hematopoietic supportive stromal cells [94] and multipotent cells able to differentiate towards the three mesenchymal lineages [89,95].

It is striking that cells isolated from adult vascular walls [76] are able not only to give rise to different mesodermal tissues, but also to sustain hematopoiesis. However, MSC derived from dental pulp, when transplanted subcutaneously did not form BM cavity [23, 92], which indicates that the property to develop hematopoietic tissue is specific of some MSC, most likely, of those derived from mesodermal progenitors surrounding the fetal aorta, and could actually be useful to identify MSC from different origins. As discussed above, vascular mural cells from the head may have a different ontogeny [78] and we can speculate that MSC from dental pulp are derived from neural crest progenitor cells that differentiate into pericytes and vSMC of the head vessels.

Since AT-MSC were shown to be very similar to BM-MSC, and that coinfusion of AT-MSC with mobilized human peripheral blood was shown to enhance human hematopoietic engraftment in immunodeficient mice [96], we investigated their capacity to support hematopoiesis. Cord blood mononuclear cells or cord blood CD34$^+$ cells magnetically selected were plated onto BM-MSC or AT-MSC monolayers and cultured for up to three weeks. Half of the medium was changed weekly and the non-adherent cells produced were analyzed by flow cytometry or cytospins stained with May-Grünwald Giemsa (MGG). The presence of myeloid progenitors was determined in clonogenic assays for granulocyte macrophage-colony forming units (GM-CFU) in presence of granulocyte macrophage colony stimulating factor (GM-CSF). Fibroblasts isolated from human skin (HSF) were used as a negative control. We observed that AT-MSC were able to sustain proliferation and differentiation of hematopoietic progenitor cells from cord blood in a very similar way to BM-MSC. BM-MSC and AT-MSC, but not HSF induced expansion of hematopoietic cells (Figure 3A) and myeloid progenitors (Figure3B). However, a tendency to induce differentiation towards macrophages was noted in co-cultures with AT-MSC. At the end of the culture, a peak of macrophage was observed in co-cultures with AT-MSC (Figure 3C) while a peak of segmented neutrophils was seen in the co-cultures with BM-MSC (Figure 3D). Since inflammatory cells, mainly macrophages, are increased in adipose tissue during obesity [97-98], and fat-cells in the bone marrow are thought to expand myeloid progenitors [99], we hypothesized that this difference could be due to the degree of differentiation of AT-MSC towards adipocytes.

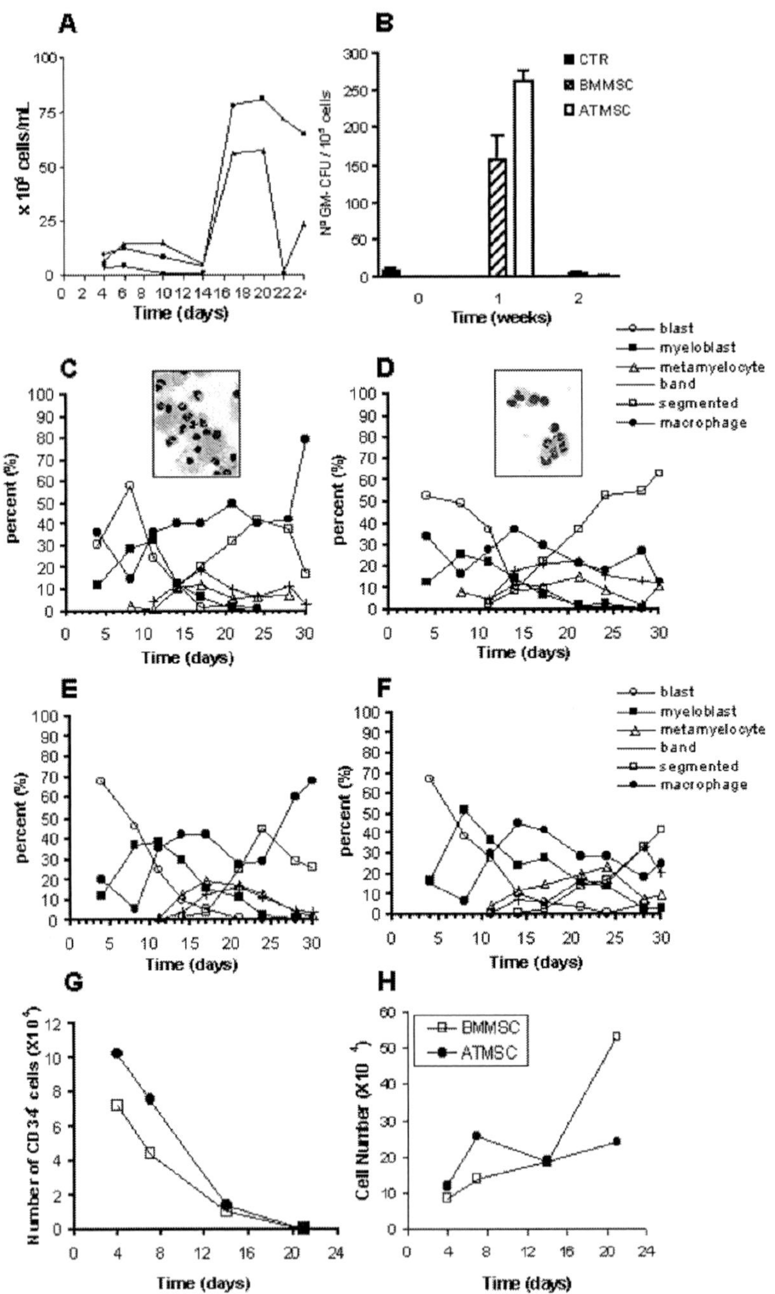

Figure 3. Adipose tissue derived mesenchymal stem cells sustain myelopoiesis. Cord blood mononuclear cells were plated onto undifferentiated BM-MSC (A-B), AT-MSC (A, C) or FHP (A) monolayers. Non-adherent cells were harvest weekly, counted and analyzed in cytospins stained with MGG. The percentage of each myeloid cell is shown. Myeloid progenitor expansion was evaluated in clonogenic assay in the presence of GM-CSF (B). The Figures are representative of three independent experiments. Inserts in C and D show the morphology of non-adherent cells at day 28, respectively. Cord blood CD34+ cells were plated onto BM-MSC or AT-MSC feeder-layers and the hematopoietic progeny was periodically counted (G) and the percentage of CD34+CD45Lo was determine by FACS (H).

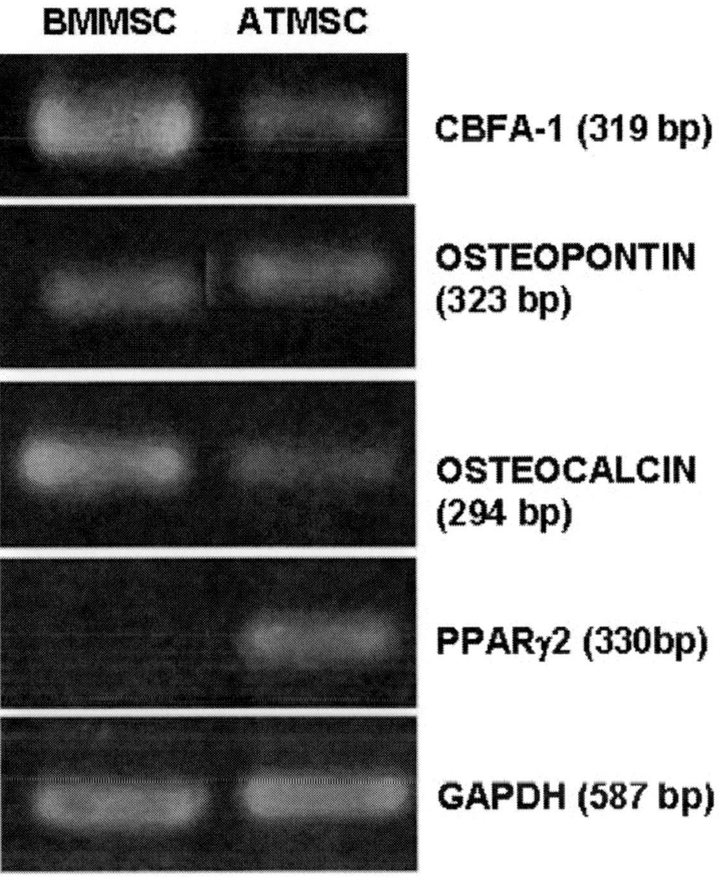

Figure 4. Expression of RUNX-2/CBFA-1, osteocalcin, osteopontin, and PPARγ2 in BM-MSC and AT-MSC. Total RNA from in vitro expanded BM-MSC and AT-MSC were isolated. The cDNA was amplified with specific primers and the RT-PCR products were analyzed in agarose gel stained with ethidium bromide. The size of the amplified products are in brackets.

This hypothesis was supported by the observation that after only one week of induction towards adipocytes, small cytoplasmic lipid drops could be observed in few cells (data not shown). Furthermore, BM-MSC constitutively expressed RUNX-2/CBFA-1, osteopontin, and osteocalcin while adipose tissue-derived MSC, although expressing RUNX-2/CBFA-1, osteopontin, and osteocalcin, also expressed PPARγ2, a marker of adipogenic differentiation [100] (Figure 4). In order to investigate this issue, AT-MSC and BM-MSC were previously cultured for one week in the presence of adipogenic medium. Supporting our hypothesis, no difference was observed in co-cultures with adipogenic induced AT-MSC (Figure 3E). However, the peak of neutrophils was not observed and a discrete increase in macrophage production was noted in the co-cultures with BM-MSC previously induced to adipocytes (Figure 3F), confirming that adipocyte differentiation is associated with rapid expansion and differentiation of myeloid progenitors into macrophages. These data raise the question of whether the increased macrophage infiltration of fat tissue during obesity [97] is due to local differentiation of circulating hematopoietic progenitors. This hypothesis is currently being investigated.

Preliminary observations in cultures seeded with magnetically selected cord blood CD34$^+$ cells confirmed these results. Although no difference in the percentage of CD34$^+$ cells in the supernatant was observed (Figure 3G), the number of hematopoietic cells produced by BM-MSC was higher (Figure 3H), confirming that AT-MSC induce mainly proliferation and differentiation of hematopoietic progenitors, while BM-MSC were able to maintain HSC.

During the preparation of this chapter, a manuscript [101] proposing that human AT-MSC sustained proliferation of hematopoietic progenitor cells but do not maintain HSC was published. Undifferentiated AT-MSC showed a very similar profile, that is, rapid differentiation of hematopoietic cells, to AT-MSC previously cultured for four weeks under adipogenic inductive conditions. This is in agreement with our observations and strengthens the hypothesis of a predominant adipocyte phenotype of AT-MSC, confirming previous reports [49].

In BM, hematopoietic progenitor cells are associated with specific niches and the observation that osteoblasts sustain HSC *in vitro* [102] led to recent reports that unequivocally showed that the subendosteal region of marrow cavity constitute a specific niche for HSC, controlling self-renewal, quiescence, proliferation and differentiation of HSC [103-105]. We then suggest that adipose tissue-derived stromal layers contain committed adipogenic precursors in a higher frequency than BM stromal layers that display a higher frequency of osteo-chondrogenic clones [8, 49], which could be responsible for the differences observed in the capacity to sustain hematopoiesis. This hypothesis is currently been investigated and although Corre and colleagues [101] have mentioned that AT-MSC differentiated into osteoblasts did not support HSC, we should point out that the maturation stage of the cells is important. Immature, but not mature, osteoblast, responded to osteotropic factors and sustained osteoclastogenesis *in vitro* [106] and *in vivo* [107]. BM-MSC cultured under osteogenic medium for more then two weeks show numerous foci of mineralization, but loses the hematopoietic supportive capacity [92, Balduíno, A., 2006, personal communication]. Actually, human BM-MSC cultured for two weeks under osteogenic inductive conditions formed ectopic bone after transplantation into the subcutaneous tissue of immunocompromised mice but not the bone marrow [92]. These results suggest that the degree of osteogenic differentiation of BM-MSC modifies their interaction with the hematopoietic cells, which means that BM niches depend upon a fine control of fate decision and differentiation pathways of these cells.

CONCLUSION

Recent findings strongly support the notion that adult MSC are potentially useful for the treatment of chronic-degenerative diseases. Although MSC-like cells can be isolated from different tissues, AT-MSC have been proposed to be an alternative to BM-MSC, since fat tissue are abundant and AT-MSC, like BM-MSC, can differentiate towards all three mesenchymal lineages, adipocytes, chondrocytes, and osteoblasts. Furthermore, AT-MSC have also the potential to give rise to a myelossuportive stroma, which is the main characteristic of BM-MSC. Although differences in phenotype, differentiation potential, and in the capacity to sustain hematopoiesis were observed, the wide distribution of MSC-like cells associated with blood vessels, that include AT-MSC and BM-MSC, not only indicates the existence of a perivascular niche for MSC in adult life, but also that MSC-like cells may be related to pericytes. Moreover, these data, together with the proposed common origin from mesodermal precursor cells located at the AGM region, and that MSC might circulate, emphasize the idea that AT-MSC and BM-MSC are related to each other and to pericytes, which implies that local environmental cues have a crucial role in their fate decision. However, it is clear that investigation should continue in order to clarify the potential of each MSC population and the relationship among them. There is no doubt that subtle differences in the phenotype, gene expression profile, and differentiation potential of the MSC isolated from different tissues exist, and it is of major importance to determine whether these differences are cell autonomous or related to the microenvironment.

ACKNOWLEDGMENTS

The authors are extremely grateful to Dr. Valéria de Mello-Coelho for suggestions and helpful discussions and Dr. Maria Eugênia Leite Duarte for their advice. They also acknowledge Dr. Marcelo Aniceto de Souza for providing adipose tissue and lipoaspirates samples, and Dr. Angelo Maiolino for providing bone marrow samples.

This work was supported by CNPq and FAPERJ grants of Brazilian Ministry of Science and Technology, Brazil.

All procedures involving human samples were approved by an Investigational Review Board at the Clementino Fraga Filho University Hospital (HUCFF), Federal University of Rio de Janeiro (UFRJ), Brazil.

REFERENCES

[1]   Weiss, L. (1976). The hematopoietic microenvironment of the bone marrow: an ultrastructural study of the stroma in rats. *Anat. Rec.,* 186, 161-184.

[2]   Bianco, P., Riminucci, M., Gronthos, S., and Robey, P. G. (2001). Bone marrow stromal stem cells: nature, biology, and potential applications. *Stem Cells,* 19, 180-192.

[3]   Friedstein, A. J., Chailakhyan, R. K., and Lalykina, K. S. (1970). The development of fibroblast colonies in monolayer cultures of guinea-pig bone marrow and spleen cells. *Cell Tissue Kinet.,* 3, 393-402.

[4]   Castro-Malaspina, H., Gay, R. E., Resnick, G., Kapoor, N., Meyers, P., Chiarieri, D., McKenzie, S., Broxmeyer, H. E., and Moore, M. A. (1980). Characterization of human bone marrow fibroblast colony-forming cells (CFU-F) and their progeny. *Blood,* 56, 289-301.

[5]   Caplan, A.I. (1991). Mesenchymal stem cell. *J. Orthop. Res.,* 9, 641-650.

[6]   Prockop D. J. (1997). Marrow stromal cells as stem cells for nonhematopoeitic tissues. *Science,* 276, 71-74.

[7]   Pittenger, M. F., Mackay, A. M., Beck, S. C., Jaiswal, R. K., Douglas, R., Mosca, J. D., Moorman, M. A., Simonetti, D. W., Craig, S., and Marshak, D. R. (1999). Multilineage potential do adult human mesenchymal stem cells. *Science,* 284, 143-147.

[8]   Muraglia, A., Cancedda, R., and Quarto, R. (2000). Clonal mesenchymal progenitors from human bone marrow differentiate in vitro according to a hierarchical model. *J. Cell Sci.*, 113, 1161-166.

[9]   Dennis, J. E. and Charbord, P. (2002). Origin and differentiation of human murine stroma. *Stem Cells*, 20, 205-214.

[10]   Gronthos, S., Zannetino, A. C., Hay, S. J., Shi, S., Graves, S. E., Kortesidis, A., and Simmons, P. J. (2003). Molecular and cellular characteristics of highly purified stromal stem cells derived from human bone marrow. *J. Cell Sci.*, 116, 1827-1835.

[11]   Toma, C., Pittneger, M. F., Cahill, K. S., Byrne, B. J., and Kessler, P. D. (2002). Human mesenchymal stem cells differentiate to a cardiomyocyte phenotype in the adult murine heart. *Circulation*, 105, 93-98.

[12]   Wang, J.S., Shum-Tim, D., Chedrawy, E., and Chiu, R. C. (2001). The coronary delivery of marrow stromal cells or myocardial regeneration: pathophysiological and therapeutic implications. *J. Thorac. Cardiovasc. Surg.*,122, 699–705.

[13]   Silva, G. V., Litovsky, S., Assad, J. A.R., Sousa, A. L.S., Martin, B. J., Vela, D., Coulter, S. C., Lin, J., Ober, J., Vaughn, W. K., Branco, R. V.C., Oliveira, E. M., He, R., Geng, Y-J., Willerson, J. T., and Perin, E. C. (2005). Mesenchymal stem cells differentiate into an endothelial phenotype, enhance vascular density, and improve heart function in a canine chronic ischemia model. *Circulation*, 111, 150-156.

[14]   Lopes, F. R. P., Campos, L. C. M., Corrêa Jr., J. D., Balduino, A., Lora, S., Langone, F., Borojevic, R., and Martinez, A. M. B. (2006). Bone marrow stromal cells and resorbable collagen guidance tubes enhance sciatic nerve regeneration in mice. *Exp. Neurol.*, 198, 457-468.

[15]   Gronthos, S. and Zannettino, A. C. W. (2006). Potential of bone marrow stromal stem cells to repair bone defects and fractures. *J. Stem Cells*, 1, 35-45.

[16]   Al-Khaldi, A., Al-Sabti, H., Galipeau, J., and Lachapelle, K. (2003). Therapeutic angiogenesis using autologous bone marrow stromal cells: Improved blood flow in a chronic limb ischemia model. *Ann. Thorac. Surg.*, 75, 204-209.

[17]   Le Blanc, K. and Ringdén, O. (2006). Mesenchymal stem cells: properties and role in clinical bone marrow transplantation. *Curr. Opin. Immunol.*, 18, 586-591.

[18]   Zuk, P. A., Zhu, M., Ashjian, P., De Ugarte, D. A., Huang, J. I., Mizuno, H., Alfonso, Z. C., Fraser, J. K., Benhaim, P., and Hedrick, M. H. (2002). Human adipose tissue is a source of multipotente stem cells. *Mol. Biol. Cell,* 12, 4279-4295.

[19]   Huss, R., Lange, C., Weissinger, E. M., Kolb, H-J., and Thalmeier, K. (2000). Evidence of peripheral blood-derived, plastic-adherent CD34–/low hematopoietic stem cell clones with mesenchymal stem cell characteristics. *Stem Cells*, 18, 252-260.

[20]   Kuznetsov, S. A., Mankani, M. H., Gronthos, S., Satomura, K., Bianco, P., and Robey, P. G. (2001). Circulating skeletal stem cells. *J. Cell Biol.*, 153, 1133-1139.

[21]   De Bari, C., Dell'Accio, F., Tvlzanowski, P., and Luyten, F. P. (2001). Multipotent mesenchymal stem cells from adult human synovial membrane. *Arthritis. Rheum.*, 44, 1928-1942.

[22]   Young, H. E., Steele, T. A., Bray, R. A., Hudson, J., Floyd, J. A., Hawkins, K., Thomas, K., Austin, T. Edwards, C. Cuzzourt, J., Duenzl, M., Lucas, P. A., and Black, A. C., Jr. (2001). Human reserve pluripotent mesenchymal stem cells are present in the connective tissues of skeletal muscle and dermis derived from fetal, adult, and geriatric donors. *Anat. Rec.*, 264, 51-62.

[23]   Shi, S. and Gronthos, S. (2003). Perivascular niche of postnatal mesenchymal stem cells in human bone marrow and dental pulp. *J. Bone Miner Res.,* 18, 696-704.

[24]   Kögler, G., Sensken, S., Airey, J. A., Trapp, T., Müschen, M., Feldhahn, N., Liedtke, S., Sorg, R. V., Fischer, J., Rosenbaum, C., Greschat, S., Knipper, A., Bender, J., Degistirici, Ö., Gao, J., Caplan, A. I., Colletti, E. J., Almeida-Porada, G., Müller, H. W., Zanjani, E., and Wernet, P. (2004). A new human somatic stem cell from placental cord blood with intrinsic pluripotent differentiation potential. *J. Exp. Med.,* 200, 123-135.

[25]   Shih, D. T., Lee, D-C., Chen, S-C., Tsai, R-Y., Huang, C-T., Tsai, C-C., Shen, E-Y., and Chiu, W-T. (2005). Isolation and characterization of neurogenic mesenchymal stem cells in human scalp tissue. *Stem Cells,* 23, 1012-1020.

[26]   Silva-Meirelles, L., Chagastelles, P. C., and Nardi, N. B. (2006). Mesenchymal stem cells reside in virtually all post-natal organs and tissues. *J. Cell Sci.,* 119, 2204-2213.

[27]   Short, B., Brouard, N., Occhiodoro-Scott, T., Ramakrishman, A., and Simmons, P. J. (2003). Mesenchymal stem cells. *Arch. Med. Res.,* 34, 565-571.

[28]   Mizuno, H., Zuk, P. A., Zhu, M., Lorenz, P., Benhaim, P., and Hedrick, M. H. (2002). Myogenic differentiation by human processed lipoaspirate cells. *Plast Reconstr. Surg.,* 109, 199-209.

[29]   Dragoo,L. J., Lieberman, J. R., Lee, R. S., Deugarte, D. A., Lee, Y., Zuk, P. A., Hedrick, M. H., and Benhaim, P. (2005). Tissue-engineered bone from BMP-2-transduced stem cells derived from human fat. *Plast Reconstr. Surg.,* 115, 1665-1673.

[30]   Winter, A., Breit, S., Parsch, D., Benz, K., Steck, E., Hauner, H., Weber, R. M., Ewerbeck, V., and Richter, W. (2003). Cartilage-like gene expression in differentiated human stem cell spheroids. *Arthritis. Rheum.,* 48, 418-429.

[31]   Rodriguez, A-M., Pisani, D., Dechesne, C. A., Turc-Carel, C., Kurzenne, J-Y., Wdziekonski, B., Villageois, A., Bagnis, C., Breittmayer, J-P., Groux, H., Ailhaud, G., and Dani, C. (2005). Transplantation of a multipotent cell population from human adipose tissue induces dystrophin expression in the immunocompetent mdx mouse. *J. Exp. Med.,* 201, 1397-1495.

[32]   Planat-Bernard, V., Silvestre, J-S., Cousin, B., André, M., Nibbelink, M., Tamarat, R., Clergue, M., Manneville, C., Saillan-Barrreau, C., Duriez, M., Tedgui, A., Levy, B., Pénicaud, L., and Casteilla, L. (2004). Plasticity of human adipose lineage cells toward endothelial cells. Physiological and therapeutic perspectives. *Circulation,* 109, 656-663.

[33]   Le Blanc, K., Rasmusson, I., Sundberg, B., Götherström, C., Hassan, M., Uzunel, M., and Ringdén, O. (2004). Treatment of severe acute graft-versus-host disease with third party haploidentical mesenchymal stem cells. *Lancet,* 636:1439-1441.

[34]   Garcia-Olmo, D., García-Arranz, M., Herreros, D., Pascual, I., Peiro, C., and Rodríguez-Montes, J. A. (2004). A phase I clinical trial of the treatment of Crohn's fistula by adipose mesenchymal stem cell transplantation. *Dis. Colon Rectum.,* 48, 1416–1423.

[35]   Yañez, R., Lamana, M. L., García-Castro, J., Colmenero, I., Ramirez, M., and Bueren, J. A. (2006). Adipose tissue-derived mesenchymal stem cells (AD-MSCS) have immunosuppressive properties applicable for the in vivo control of the graft-versus-host disease (GVHD). *Stem Cells,* 24, 2582-2591.

[36]   Beresford, J. N., Benneth, J. H., Devlin, C., Leboy, P. S., and Owen M. E. (1992). Evidence for an inverse relationship between the differentiation of adipocytic and osteogenic cells in rat marrow stromal cell cultures. *J. Cell Sci.,* 102, 341-351.

[37] Nuttall, M. E. and Gimble, J. M. (2004). Controlling the balance between osteoblastogenesis and adipogenesis and the consequent therapeutic implications. *Curr. Op. Pharmacol.*, 4, 290-294.

[38] Kaplan, F. S. (1996). Skin and bones. *Arch. Dermatol.*, 132, 815-818.

[39] Shore, E. M., Ahn, J., de Beur, S. J., Li, M., Xu, M., Gardner, R. J. M., Zasloff, M. A., Whyte, M. P., Levine, M. A., and Kaplan, F. S. (2002). Paternally inherited inactivating mutations of the GNAS1 gene in progressive osseous heteroplasia. *N. Engl. J. Med.*, 346, 99-106.

[40] Wagner, W., Wein, F., Seckinger, A., Frankhauser, M., Wirkner, U., Krause, U., Blake, J., Schwager, C., Eckstein, V., Ansorge, W., and Ho, H. (2005). Comparative characteristics of mesenchymal stem cells from human bone marrow, adipose tissue, and umbilical cord blood. *Exp. Hematol.*, 33, 1402-1416.

[41] Shahdadfar, A., Frønsdal, K., Haug, T., Reinholt, F. P., and Brinchmann, J. E. (2005). In vitro expansion of human mesenchymal stem cells: choice of serum Is a determinant of cell proliferation, differentiation, gene expression, and transcriptome stability. *Stem Cells*, 23, 1357-1366.

[42] Chen, C. S., Mrksich, M., Huang, S., Whitesides, G.M., and Ingber. D. E. (1997). Geometric control of cell life and death. *Science*, 276, 1425-1428.

[43] Lelièvre, S. A., Weaver, V. M., Nickerson, J. A., Larabell, C. A., Bhaumik, A., Petersen, O. W., and Bissel, M. J. (1998). Tissue phenotype depends on reciprocal interactions between the extracellular matrix and the structural organization of the nucleus. *Proc. Nat. Acad. Sci. USA,* 95, 14711-14716.

[44] Thomas, C. H., Collier, J. H., Sfeir, C. S., and Healy, K. E. (2002). Engineering gene expression and protein synthesis by modulation of nuclear shape. *Proc. Nat. Acad. Sci. USA,* 99, 1972-1977.

[45] McBeath, R., Pirone, D. M., Nelson, C. M., Bhadriraju, K., and Chen, C. S. (2004). Cell shape, cytoskeletal tension, and RhoA regulate stem cell lineage commitment. *Developmental Cell*, 6, 483-495.

[46] Rossi, M. I. D., Barros, A. P. D. N., Baptista, L. S., Garzoni, L. R., Meirelles, M. N., Takiya, C. M., Pascarelli, B. M., Dutra, H. S., and Borojevic, R. (2005). Multicellular spheroids of bone marrow stromal cells: a three-dimensional in vitro culture system for the study of hematopoietic cell migration. *Braz. J. Med. Biol. Res.,* 38, 455-462.

[47] Chun, T-H., Hotary, K. B., Sabeh, F., Saltiel, A. R., Allen, E. D., and Weiss, S. J. (2006). A pericellular collagenase directs the 3-Dimensional development of white adipose tissue. *Cell*, 125, 577-591.

[48] Dicker, A., Le Blanc, L., Aström, G., van Harmelen, V., Götherström, C., Blomqvist, L., Arner, P., and Rydén, M. (2005). Functional studies of mesenchymal stem cells derived from adult human adipose tissue. *Exp. Cell Res.*, 308, 283-290.

[49] Liu, T. M., Martina, M., Hutmacher, D. W., Hoi, J., Hui, P., Lee, E. H., and Lim, B. (2006). Identification of common pathways mediating differentiation of bone marrow and adipose tissues derived human mesenchymal stem cells (MSCs) into three mesenchymal lineages. *Stem Cells*, published online doi:10.1634/stemcells.2006-0394.

[50] Im, G-I., Shin, Y-W., and Lee, K-B. (2005). Do adipose tissue-derived mesenchymal stem cells have the same osteogenic and chondrogenic potential as bone marrow-derived cells? *Osteoarthritis Cartilage*, 13, 845-853.

[51] Conejero, J. A., Lee, J. A., Parrett, B. M., Terry, M., Wear-Maggitti, K., Grant, R. T., and Breitbart, A. S. (2006). Repair of palatal bone defects using osteogenically differentiated fat-derived stem cells. *Plast Reconstr. Surg.*, 117, 857-863.

[52] Hicok, K. C., du Laney, T. V., Zhou, Y. S., Halvorsen, Y-D. C., Hitt, D. C., Cooper, L. F., and Gimble, J. M. (2004). Human adipose-derived adult stem cells produce osteoid in vivo. *Tissue Eng.*, 10, 371-380.

[53] Dominici, M. Le Blanc K., Müeller, I., Slaper-Cortenbach, I., Marini, F., Krause, D., Deans, R., Keating, A., Procokop, D. J., and Horwitz, E. (2006). Minimal criteria for defining multipotent mesenchymal stromal cells. The International Society for Cellular Therapy position statement. *Cytotherapy*, 8, 315-317.

[54] Simmons, P. J. and Torok-Storb, B. (1991). Identification of stromal cell precursors in human bone marrow by a novel monoclonal antibody, STRO-1. *Blood*, 78, 55-62.

[55] De Ugarte, D. A., Alfonso, Z., Zuk, P. A., Elbarbary, A., Zhu, M., Ashjian, P., Benhaim, P., Hedrick, M. H., and Fraser, J. K. (2003). Differential expression of stem cell mobilization-associated molecules on multi-lineage cells from adipose tissue and bone marrow. *Immunol. Lett.*, 89, 267- 270.

[56] Minguell, J. J., Erices, A., and Conget, P. (2001). Mesenchymal stem cells. *Exp Biol Med*, 226, 507–520.

[57] Gronthos, S., Franklin, D. M., Leddy, H. A., Robey, P. G., Storms, R. W., Gimble, J. M. (2001). Surface protein characterization of human adipose tissue-derived stromal cells. *J. Cell Physiol.*, 189, 54-63.

[58] Mitchell, J. B., McIntosh, K., Zvonic, S., garret, S., Floyd, E., Kloster, A., Di Halvorsen, Y., Storms, R. W., Goh, B., Kilroy, G., Wu, X., and Gimble, J. M. (2006). Immunophenotype of human adipose-derived cells: temporal changes in stromal-associated and stem cell-associated markers. *Stem Cells*, 24, 376-385.

[59] Boquest, A. C., Shahdadfar A., Fronsdal, K., Sigurjonsson O., Tunheim, S. H., Collas, P., and Brinchmann, J. E. (2005). Isolation and transcription profiling of purified uncultured human stromal stem cells: Alteration of gene expression after in vitro cell culture. *Mol. Biol. Cell*, 16, 1131-1141.

[60] Colter, D. C., Class, R., DiGiolamo, C. M., and Prockop, D. J. (2000). Rapid expansion of recycling stem cells in cultures of plastic-adherent cells from human bone marrow. *Proc. Natl. Acad. Sci. USA*, 97, 3213-3218.

[61] Zengin, E., Chalajour, F., Gehling, U. M., Ito, W. D., Treede, H., Lauke, H., Weil, J., Reichenspurner, H., Kilic, N., and Ergün, S. (2006). Vascular wall resident progenitor cells: a source for postnatal vasculogenesis. *Development*, 133, 1543-1551.

[62] Godin, I. and Cumano, A. (2002). The hare and the tortoise: an embryonic hematopoietic race. *Nat. Rev. Immunol.*, 2, 593-604.

[63] Charbord, P., Tavian, M., Humeau, L., and Péault, B. (1996). Early ontogeny of the human marrow from long bones: An immunohistochemical study of hematopoiesis and its microenvironment. *Blood,* 87, 4109-4119.

[64] Arai, F., Ohneda, O., Miyamoto, T., Zhang, X. Q., and Suda, T. (2002). Mesenchymal stem cells in perichondrium express activated leukocyte cell adhesion molecule and participate in bone marrow formation. *J. Exp. Med.,* 195, 1543-1563.

[65] Bianco, P. and Cossu, G. (1999). Uno, nessuno e centomila: searching for the identity of mesodermal progenitors. *Exp. Cell Res.*, 251, 257-263.

[66] Cossu, G. and Bianco, P. (2003). Mesoangioblasts – vascular progenitors for extravascular mesodermal tissues. *Curr. Opin. Genet Dev.*, 13, 537-542.

[67] Diaz-Flores, L., Gutierrez, R., Lopez-Alonso, A., Gonzalez, R., and Varela, H. (1992). Pericytes as a supplementary source of osteoblasts in periosteal osteogenesis. *Clin. Orthop. Relat. Res.*, 75, 280-286.

[68] Gregoire, F. M., Smas, C. M., and Sul, H. S. (1998). Understanding adipocyte differentiation. *Physiol. Rev.*, 78, 783-809.

[69] Otto, T. C. and Lane, M. D. (2005). Adipose development: From stem cell to adipocyte. *Crit. Rev. Biochem. Mol. Biol.*, 40, 229-242.

[70] Kawaguchi, N., Toriyama, K., Nicodemous-Lena, E., Inou, K., Torri, S., and Kitagawa, Y. (1998). De novo adipogenesis in mice at the site of injection of basement membrane and basic fibroblast growth factor. *Proc. Natl. Acad. Sci. USA*, 95, 1062-1066.

[71] Iyama, K., Ohzono, K., and Usuku, G. (1979). Electron microscopical studies on the genesis of white adipocytes: differentiation of immature pericytes into adipocytes in transplanted preadipose tissues. *Virchows Arch. B Cell Pathol. Incl. Mol. Pathol.*, 31, 143-1455.

[72] Richardson, R. L., Hausman, G. J., and Campion, D. R. (1982). Response of pericytes to thermal lesion in the inguinal fat pad of 10-day-old rats. *Acta Anat. (Basel)*, 114, 41-57.

[73] Gerhardt, H. and Betsholtz, C. (2003). Endothelial-pericyte interactions in angiogenesis. *Cell Tissue Reg.,* 314, 15-23.

[74] Armulik, A., Abramsson, A., and Betsholtz, C. (2005). Endothelial/pericytes interactions. *Circ. Res.*, 97, 512-523.

[75] Doherty, M. J., Ashton, B. A., Walsh, S., Beresford, J. N., Grant, M. E., and Canfield, A. E. (1998). Vascular pericytes express osteogenic potential in vitro and in vivo. *J. Bone Miner Res.*, 13, 828-838.

[76] Tintut, Y., Alfonso, Z., Saini, T., Radcliff, K., Watson, K., Boström, K., and Demer, L. L. (2003). Multilineage potential of cells from the artery wall. *Circulation,* 108, 2505-2510.

[77] Farrington-Rock, C., Crofts, N. J., Doherty, M. J., Ashton, B. A., Griffin-Jones, C., and Canfield, A. E. (2004). Chondrogenic and adipogenic potential of microvascular pericytes. *Circulation*, 110: 2226-2232.

[78] Etchevers, H. C., Vincent, C., Le Douarin, N. M., and Couly, G. F. (2001). The cephalic neural crest provides pericytes and smooth muscle cells to all blood vessels of the face and forebrain. *Development*, 128, 1059-1068.

[79] Mikawa, T. and Gourdie, R. G. (1996). Pericardial mesoderm generates a population of coronary smooth muscle cells migrating into the heart along with ingrowth of the epicardial organ. *Int. Dev. Biol.*, 174, 221-232.

[80] DeRuiter, M. C., Poelmann, R. E., VanMunsteren, J. C., Mironov, V., Markwald, R. R., and Gittenberger-de Groot, A.C. (1997). Embryonic endothelial cells transdifferentiate into mesenchymal cells expressing smooth muscle actins in vivo and in vitro. *Circ. Res.*, 80, 444-451.

[81] Drake, D. J., Hungerford, J. E., and Little, C. D. (1998). Morphogenesis of the first blood vessels. *Ann. N. Y. Acad. Sci.*, 857, 155-179.

[82] Hungerford, J. E. and Little, C. D. (1999). Developmental biology of the vascular smooth muscle cell: building a multilayered vessel wall. *J. Vasc. Res.*, 36, 2-27.

[83] Minasi, M. G., Riminucci, M., De Angelis, L., Borello, U., Berarducci, B., Innocenzi, A., Caprioli, A., Sirabella, D., Baiocchi, M., De Maria, R., Boratto, R., Jaffredo, T., Broccoli, V., Bianco, P., and Cossu, G. (2002). The meso-angioblast: a multipotent, self-renewing cell that originates from the dorsal aorta and differentiates into most mesodermal tissues. *Development*, 129, 2773-2783.

[84] Pouget, C., Gautier, R., Teillet, M-A., and Jaffredo, T. (2006). Somite-derived cells replace ventral aortic hemangioblasts and provide aortic smooth muscle cells of the trunk. *Development*, 133, 1013-1022.

[85] Lamagna, C. and Bergers, G. (2006). The bone marrow constitutes a reservoir of pericytes progenitors. *J. Leukoc. Biol.*, 80, 677-681.

[86] Erices, A., Conget, P., and Miguell, J. J. (2000). Mesenchymal progenitor cells in human umbilical cord blood. *Br. J. Haematol.*, 109, 235-242.

[87] Campagnoli, C., Roberts, I. A. G., Kumar, S., Bennett, P. R., Bellantuono, I., and Fisk, N. M. (2001). Identification of mesenchymal stem/progenitor cells in human first-trimester fetal blood, liver, and bone marrow. *Blood*, 98, 2396-2402.

[88] O'Donoghue, K, Chan, J., Fuente, J. de la, Kennea, N., Sandison, A., Anderson J. R., Roberts, I. A. G., and Fisk, N. M. (2004). Microchimerism in female bone marrow and bone decades after fetal mesenchymal stem-cell trafficking in pregnancy. *Lancet*, 364, 179-182.

[89] Mendes, S. C., Robin, C., and Dzierzak, E. (2004). Mesenchymal progenitor cells localize within hematopoietic sites throughout ontogeny. *Development*, 132, 1127-1136.

[90] Hong, K. M., Burdick, M. D., Phillips, R. J., Heber, D., and Strieter, R. M. (2005) Characterization of human fibrocytes as circulating adipocyte progenitors and the formation of human adipose tissue in SCID mice. *FASEB J.*, 19, 2029-2031.

[91] Crossno, J. T. Jr., Majka, S. M., Grazia,T., Gill, R. G., and Klemm, D. J. (2006). Rosiglitazone promotes development of a novel adipocyte population from bone marrow–derived circulating progenitor cells. *J. Clin. Invest.*, 116,320-328.

[92] Miura, Y., Gao, Z., Miura, M., Seo, B-M., Sonoyama, W., Chen, W., Gronthos, S., Zhang, L., and Shi, S. (2006). Mesenchymal stem cell-organized bone marrow elements: An alternative hematopoietic progenitor resource. *Stem Cells*, 24, 2428-2436.

[93] Muguruma, Y., Yahata, T., Miyatake, H., Sato, T., Uno, T., Itoh, J., Kato, S., Ito, M., Hotta, T., and Ando, K. (2006). Reconstitution of the functional human hematopoietic microenvironment derived from human mesenchymal stem cells in the murine bone marrow compartment. *Blood*, 107, 1878-1887.

[94] Oostendorp, R. A., Harvey, K. N., Kusadasi, N., de Bruijn, M. F. T. R., Saris, C., Ploemacher, R. E., Medvinsky, A. L., and Dzierzak, E. (2002). Stromal cell lines from mouse aorta-gonads-mesonephros subregions are potent supporters of hematopoietic stem cell activity. *Blood*, 99, 1183-1189.

[95] Duran, C., Robin, C, and Dzierzak, E. (2006). Mesenchymal lineage potentials of aorta-gonad-mesonephros stromal clones. *Haematologica*, 91, 1172-1179.

[96] Kim, J. S., Cho, H. H., Kim, Y. J., Seo, S. Y., Kim, H. N., Lee, J. B., Kim, J. H., Chung, J. S., and Jung, J. S. (2005). Human adipose stromal cells expanded in human serum promote engraftment of human peripheral blood hematopoietic stem cells in NOD/SCID mice. *Biochem. Biophys. Res. Commun.*, 329, 25-31.

[97]   Weisberg, S. P., McCann, D., Desai, M., Rosenbaum, M., Leibel, R. L., and Ferrante, A. W. Jr. (2003). Obesity is associated with macrophage accumulation in adipose tissue. *J. Clin. Invest.*, 112, 1798-1808.

[98]   Xu, H., Barnes, G. T., Yang, O., Tan, G., Yang, D., Chou, C. J., Sole, J., Nichols, A., Ross, J. S., Tartaglia, L. A., and Chen, H. (2003). Chronic inflammation in fat plays a crucial role in the development of obesity-related insulin resistance. *J. Clin. Invest.*, 112, 1821-1830.

[99]   Corre, J., Planat-Benard, V., Corberand, J. X., Pénicaud, L., Casteilla, L., and Laharrague, P. (2004). Human bone marrow adipocytes support complete myeloid and lymphoid differentiation from human CD34^{+} cells. *Br. J. Haematol.*, 127, 344-347.

[100] Ren, D., Collingwood, T. N., Rebar, E. J., Wolffe, A. P., and Camp, H. S. (2002). PPARgamma knockdown by engineered transcription factors: exogeneous PPARgamma2 but not PPARgamma1 reactivates adipogenesis. *Genes Develop.*, 16, 27-32.

[101] Corre, J., Barreau, C., Cousin, B., Chavoin, J-P., Caton, D., Fournial, G., Penicaud, L., Castetilla, L., and Laharrague, P. (2006). Human subcutaneous adipose cells support complete differentiation but not self-renewal of hematopoietic progenitors. *J. Cell Physiol.*, 208, 282-288.

[102] Taichman, R. S., Reilly, M. J., and Emerson, S. G. (1996) Human osteoblasts support human hematopoietic progenitor cells in vitro bone marrow cultures. *Blood,* 87, 518–524.

[103] Calvi, L. M., Adams, G. B., Wlbrecht, K. W., Weber, J. M., Olson, D. P., Knight, M. C,, Martin, R. P., Schipani, E., Divieti, P., Bringhurst, F. R., Milner, L. A., Kronenberg, H. M., and Scadden, D. T. (2003). Osteoblastic cells regulate the haematopoietic stem cell niche. *Nature,* 425, 841-846.

[104] Zhang, J., Niu, C., Ye, L., Huang, H., He, X., Tong, W-G., Rossi, J., Haug, J., Johnson, T., Feng, J. Q., Harris, S., Wiedemann, L. M., Mishina, Y., and Li, L. (2003)Identification of the haematopoietic stem cell niche and control of the niche size. *Nature,* 425, 836-841.

[105] Balduino, A., Hurtado, S. P., Frazão, P., Takiya, C. M., Alves, L. M., Nasciutti, L-E., El-Cheikh, M. C., and Borojevic, R. (2005). Bone marrow subendosteal microenvironment harbours functionally distinct haemosupportive stromal cell populations. *Cell Tissue Res,* 319, 255–266.

[106] Atkins, G. J., Kostakis, P., Pan, B., Farrugia, A., Gronthos, S., Evdokiou, A., Harrison, K., Findlay, D. M., and Zannettino, A. C. (2003). RANKL expression is related to the differentiation state of human osteoblasts. *J. Bone Miner Res.*, 18, 1088-1098.

[107] Corral, D. A., Amling, M., Priemel, M., Loyer, E., Fuchs, S., Ducy, P., Baron, R., and Karsenty, G. (1998). Dissociation between bone resorption and bone formation in osteopenic transgenic mice. *Proc Natl. Acad. Sci. USA,* 95, 13835-13840.

In: Stem Cell Research Progress
Editor: Prasad S. Koka, pp. 117-135

ISBN: 978-1-60456-308-5
© 2008 Nova Science Publishers, Inc.

Chapter 8

FROM THE BONE MARROW TO THE PERIPHERY: THE ONTOGENY AND FUNCTION OF DENDRITIC CELL POPULATIONS

A. I. Proietto[1,2] and Li Wu[1]

[1] Immunology Division, The Walter and Eliza Hall Institute,
1G Royal Parade, Parkville, Victoria 3050, Australia
[2] Department of Medical Biology, The University of Melbourne.
Royal Parade, Parkville, Victoria 3010, Australia

ABSTRACT

Dendritic cells (DC) are a heterogenous population of bone marrow-derived cells residing in primary and secondary lymphoid organs. All DC share a common ability to process and present antigen to naïve T cells for the initiation of an immune response. However, they differ in surface markers, migratory patterns, localization and cytokine production. DC had been considered as myeloid cells for some time. However, recent findings have demonstrated that DC can develop not only from myeloid committed progenitors, but also from early lymphoid committed progenitors. The ability of DC to develop from progenitors committed to different lineages correlates with the surface expression of Flt3 receptor. The developmental events downstream from the lymphoid and myeloid committed progenitors as well as the molecular mechanisms required for DC development have been less well characterized. Recent studies have provided evidence for an intra-splenic precursor population that gives rise to the DC subsets *in situ*. The impaired DC development found in mice deficient in certain transcription factors known to be required for hemopoiesis has provided clues to the important role of transcription factors in regulating DC development. The recent finding that steady-state "dendropoiesis" can be recreated *in vitro* using bone marrow progenitors in the presence of Flt3 ligand will facilitate further studies of DC development nand function. This review summarizes recent crucial findings on the development of a functional DC system.

INTRODUCTION

Dendritic Cells (DC) represent a sparsely distributed population of bone marrow-derived cells that play a central role in the induction and maintenance of an immune response [1]. Residing in many organs of the body including non-lymphoid, such as the skin and mucosae, and lymphoid organs, such as the thymus, spleen and lymph nodes, DC in mouse can be identified as conventional antigen presenting DC (cDC) that express high levels of CD11c and MHC Class-II, and plasmacytoid DC (pDC) that are $CD45RA^+CD11c^{int}$ and MHC Class-II^{lo} [2]. cDC function as the sentinels of the immune system by picking up and processing exogenous antigen for presentation to and activation of naïve T cells in secondary lymphoid organs. In addition to inducing immune responses to foreign antigens, cDC, paradoxically are able to induce and maintain tolerance to self-antigens. Self-tolerance is a continual process occurring in the steady state and takes place both centrally and peripherally. Central tolerance occurs in the thymus where thymic cDC are involved in the deletion of self-reactive thymocytes. Self-reactive T cells not deleted in the thymus are tolerised in the periphery either through the process of anergy, which renders the T cell incapable of further activation or through deletion. Defining the conditions and factors that direct cDC toward immunity or tolerance remains one of the greatest conundrums in DC biology. In contrast to cDC, pDC are not classical antigen presenting cells in terms of their surface phenotype, morphology and function. However, upon activation, pDC can produce large amount of type-I interferon and assume the phenotype and morphology of a cDC [3], suggesting a role of pDC in controlling various viral infections.

The development and function of the DC subsets are the two areas of DC biology under intense investigation. The heterogeneity and sparsity of DC has made these studies a difficult task. However, significant progress has been made recently using both mouse and human models. This review will summarize some recent findings focussing on studies using mouse models.

DC DIVERSITY IN LYMPHOID ORGANS

DC have been described in the skin and in lymphoid organs, namely the thymus, the spleen, the lymph nodes (LNs) and Peyer's patches [Table 1]. They represent a population of cells that are heterogenous in surface phenotype. This phenotypic diversity has been correlated with functional differences between the DC subsets (described in this review).

The cDC in these organs display differential surface expression of a range of antigen markers. Several of these markers are used to segregate the different DC subtypes. While all DC express CD11c (the integrin alpha chain), MHC Class-II and low levels of the co-stimulatory molecules CD80 and CD86, they may be segregated on the basis of CD4 and $CD8\alpha$ expression [4]. Other markers useful for segregating DC subtypes include the myeloid marker CD11b, CD205 – a C-type lectin endocytic receptor, and more recently, the signal regulatory protein $Sirp\alpha$ [5].

Langerhan's cells (LCs) represent a population of cDC found in the epidermis, bronchi and mucosae [1]. They express the Langerin antigen and contain Birbeck granules – two distinguishing features of this cell type [8]. They are specialized for antigen capture in these areas and upon activation, migrate to draining LNs [7]. Dermal DC represent a second population of cDC in the skin. These also migrate to draining LNs upon activation.

The thymus – the centre for thymocyte differentiation and tolerance induction contains two cDC subsets that can be segregated on the basis of CD8α [4] and Sirp-α . When used in conjunction they can clearly segregate the thymic cDC into CD8α^+Sirp-α^- and CD8$\alpha^{-/lo}$Sirp-α^{-+} subsets [5] [Table 1] (Figure 1). Three cDC subsets have been identified in the mouse spleen based on the surface expression of CD8α, CD4 and Sirp-α. These are CD4$^-$CD8$^+$Sirp-α^-, CD4$^-$CD8$^-$Sirp-α^+ and CD4$^+$CD8$^-$Sirp-α^+ [Table 1] (Figure1). Cell turnover analysis of these populations demonstrate the three cDC subsets to be products of separate developmental lineages rather than different maturation states of the one lineage [9].

The DC populations found in the LNs are more complex. In addition to the three cDC populations of the spleen, two additional subpopulations have been described [Table 1]. These correspond to the mature CD8^{lo}CD205^{int} and CD8^{lo}CD205^{hi} cDC that migrate from the dermis and epidermis respectively, to the LNs. Subcutanous LNs contain a higher percentage of the CD8^{lo}CD205^{hi} Langerhan cell-like cells than the mesenteric LNs [7].

Table 1. Murine conventional DC (CD11c$^+$CD45RA$^-$) subtypes

Subtype	Phenotype	Percentage of total DCs in organs	
Skin [2]			
Langerhan Cells CD4$^-$CD8$^-$	CD11b$^+$, CD205^{hi}, Langerin$^+$	Not determined	
Dermal DC CD4$^-$CD8$^-$	CD11b$^+$, CD205^{int}, Langerin$^-$	Not determined	
Thymus [5, 6]			
CD4$^-$CD8$^+$	CD11b$^-$, CD205$^+$, Sirp-α^-	70%	
CD4$^-$CD8$^-$	CD11b$^+$, CD205$^+$, Sirp-α^+	30%	
Spleen [6]			
CD4$^-$CD8$^-$	CD11b$^+$, CD205$^-$, Sirp-α^+	20%	
CD4$^+$CD8$^-$	CD11b$^+$, CD205$^-$, Sirp-α^+	60%	
CD4$^-$CD8$^+$	CD11b$^-$, CD205$^+$, Sirp-α^-	20%	
Lymph Node [2, 7]		Mesenteric	Subcutaneous
CD4$^-$CD8$^-$	CD11b$^+$, CD205^{lo}, Langerin$^-$	37%	17%
	CD11b$^+$, CD205^{hi}, Langerin$^+$	4%	33%
	CD11b$^+$, CD205^{int}, Langerin$^-$	26%	20%
CD4$^+$CD8$^-$	CD11b$^+$, CD205^{lo}	4%	4%
CD4$^-$CD8$^+$	CD11b$^-$, CD205$^+$	19%	17%
Peyer's Patches [8]			
CD8$^-$	CD11b$^+$, CD205$^-$	70%	
CD8$^+$	CD11b$^-$, CD205^{hi}	10%	
CD8$^-$	CD11b$^-$, CD205$^-$	20%	

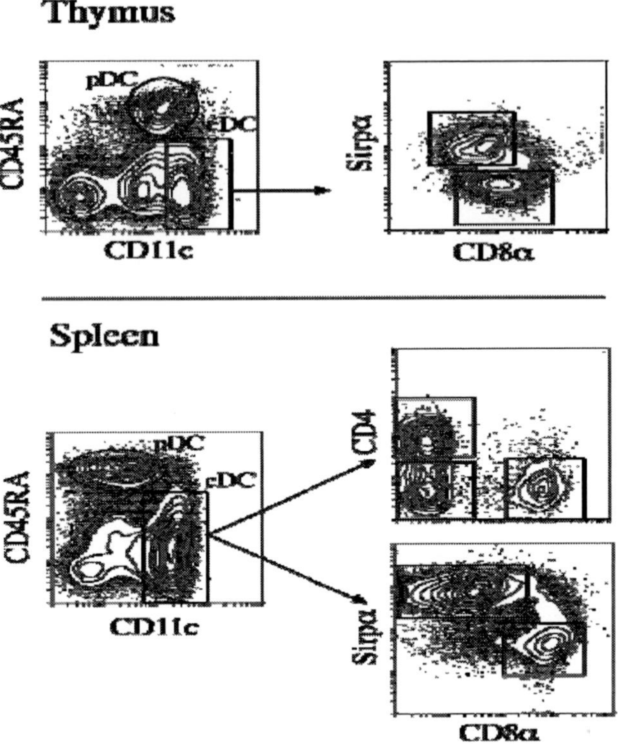

Figure 1. Thymic and splenic DC can be segregated into plasmacytoid DC (CD11c^{int}CD45RA^{hi}) and conventional DC (CD11c^{hi}CD45RA^{lo}). Thymic cDC can be further segregated into two subsets based on the expression of CD8α and Sirpα; CD8^{lo}Sirpα^{+} and CD8^{hi}Sirpα^{-}. Splenic cDC can be segregated into three subsets based on the expression of CD4 and CD8; CD4^{-}CD8^{-} (DN), CD4^{+}8^{-} and CD4^{-}8^{+} or into two subsets based on the expression of CD8α and Sirpα; CD8^{lo}Sirpα^{+} and CD8^{hi}Sirpα^{-}.

Peyer's patches are lymphoid tissues representing the site for the induction of mucosal immune responses [10]. Three populations of cDC reside in distinct anatomical locations in the Peyer's patches. The CD8^{-}CD11b^{+} DCs are found in the sub-epithelial dome region where they are positioned to capture antigens, while the CD8^{+}CD11b^{-} DC reside in T cell-rich interfollicular regions where naïve T cells may be activated. The third population described as CD8^{-}CD11b^{-} is found in both sites [10].

In addition to the cDC described above, another cell population defined as CD11c^{int}CD45RA^{+} and named plasmacytoid DC (pDC) due to their morphological resemblance to plasma cells, is also present in all lymphoid organs and in blood [11]. The fresh pDC do not have the phenotypic and functional features of the antigen presenting cDC, but can assume a cDC morphology and upregulate the cDC markers CD11c and MHC Class-II after activation with exogenous stimuli. They represent the major cell type that produce large amounts of type-1 interferon – a cytokine involved in innate immunity to virus [3, 12-15].

ONTOGENY OF DC IN THE EARLY STAGES OF MOUSE DEVELOPMENT

Neonatal mice have an immature immune system and are more susceptible to microbial infections. T cells from neonatal mice have been shown to be functionally similar to adult T cells, and thus the maturation status of DC has been implicated in affecting neonatal immunity.

A recent study showed that DC appear in the mouse spleen appear from day 1 after birth [16]. The absolute numbers and proportion of both cDC and pDC did not reach adult levels until 5wks of age. Interestingly, during this period, the composition of the cDC populations changed markedly [Table 2]. At day 1, there was a minor population of DC that were CD8$^-$CD205$^-$. The dominant population was CD8$^-$ but expressed CD205 (CD8$^-$CD205$^+$), in contrast to the CD8$^-$ population in adults which are low for CD205. Moreover, this CD8$^-$CD205$^+$ DC subset showed the capacity to produce IL12p70 – a cytokine normally produced by the CD8$^+$CD205$^+$ population in adult spleen.

Significantly, this population was no longer detectable at 3wks of age. These findings suggested that the neonatal CD8$^-$CD205$^+$ DC represent an immature stage of the CD8$^+$CD205$^+$ DC. By 2wks of age, the CD8$^-$CD205$^+$ and the CD8$^-$CD205$^-$ DC represented minor populations while the CD8$^+$CD205$^+$ became the major cDC population. However, after this time, the CD8$^-$CD205$^+$ cDC were undetectable and the CD8$^-$CD205$^-$ cDC became the major DC population. This result demonstrated that different DC populations have different developmental kinetics during ontogeny.

A similar picture of DC development during ontogeny was observed in the thymus [Table 2]. Thymic DC were detected at E17, coinciding with the emergence of CD4$^+$8$^+$ thymocytes and the beginning of thymocyte selection processes. The cDC were CD8$^-$, with the emergence of CD8$^+$ DC observed at 1wk of age. By 2wks, the CD8$^+$ population became the major population, with the relative proportions of each subset similar to those found in the adult thymus [16]. Furthermore, the pDC appeared at the same time as cDC in both thymus and spleen and their number increased with age and in parallel with the increase in the number of cDC.

Consistent with the developmental kinetics of DC during mouse ontogeny, functional analysis of neonatal DC have shown that they are immature in a number of different functions including cytokine production and antigen presentation[17-21].

Table 2. The phenotype and proportion of cDC subsets in the spleen and thymus of developing mice

Spleen

	Day 1	Week 1	Week 2	Week 3	Week 5
CD8$^-$CD205$^-$	20%	20%	20%	75%	75%
CD8$^-$CD205$^+$	80%	30%	30%	0 %	0 %
CD8$^+$CD205$^+$	0%	50%	50%	25%	25%

Thymus

	Day 1	Week 1	Week 2	Week 3	Week 5
CD8$^-$CD205$^+$	100%	70%	30%	30%	30%
CD8$^+$CD205$^+$	0%	30%	70%	70%	70%

Development of DC

Hemopoiesis represents a process whereby all blood cells are produced from a multipotent hemopoietic stem cell (HSC) located in the adult mouse bone marrow (BM) [22-25]. The blood cells can be categorized into myeloid or lymphoid cells. Myeloid cells include monocytes, macrophages, neutrophils, eosinophils, basophils, megakaryocytes and erythrocytes [26]. Lymphoid cells include T and B lymphocytes, and natural killer (NK) cells [27]. The commitment to either the lymphoid or myeloid lineage is made early on with the HSC giving rise to the committed lymphoid and myeloid progenitor cells from which all respective cell lineages are produced (Figure 2). Although the origins of both myeloid and lymphoid cells have been well characterized, the exact lineage origins of DC remains an issue that requires further clarification. Increasing evidence suggest that DC arise from both the myeloid and lymphoid lineages, representing the only cell type whose origins are not restricted to one lineage [28-30].

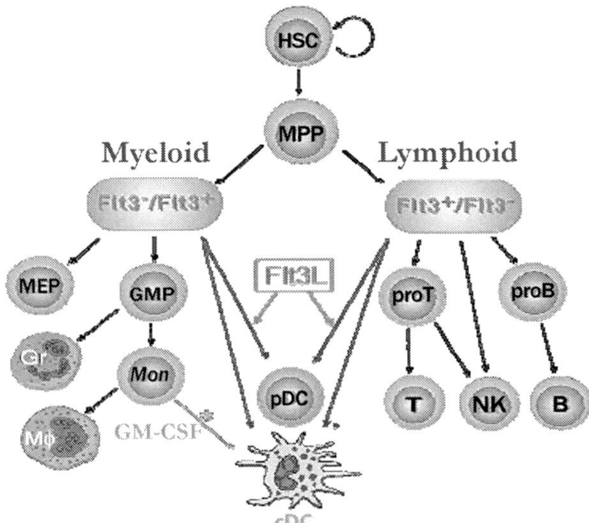

Figure 2. A developmental scheme of DC development from BM hemopoietic stem cells and lineage committed progenitors. Both steady-state cDC and pDC can be generated from the Flt3 expressing early myeloid or lymphoid progenitors. Under inflammatory conditions, i.e. in the presence of GM-CSF, monocytes can differentiate into cDC. HSC: hemopoietic stem cell; MPP: multi-potent progenitors; GMP: granulocyte and macrophage precursors; MEP: megakaryocyte and erythrocyte precursors.

Myeloid Origin of DC

The original studies on DC lineage suggested a myeloid origin for DC. Mouse BM myeloid precursors demonstrated a capacity to produce macrophages, granulocytes and DC in the presence of Granulocyte-Macrophage colony stimulating factor (GM-CSF) [31]. These findings were extended to human studies where a CD34[+] BM derived precursor differentiated into CD1a⁻ monocytes [32], a granulocyte precursor population and a bi-potential precursor population with the ability to produce mature DC when cultured in the presence of GM-CSF and TNF-α, or macrophages when cultured in the presence of M-CSF [33, 34].

Further evidence for a myeloid origin of DC came from the finding that monocytes differentiated into DC in the presence of GM-CSF and IL-4, *in vitro* [35]. This differentiation process also occured *in vitro* when monocytes reverse transmigrated in an ablumenal-to-lumenal direction across endothelial cells [36]. A similar phenomenon has been observed *in vivo* with subcutaneous tissue monocytes [36]. In this study, activated monocytes retained in the subcutaneous tissue developed into macrophages whereas those that migrated to the LNs differentiated into DC. Monocytes were also shown to differentiate into LCs in the presence of TGF-β . This cytokine is expressed at high levels by epithelial cells and thus it was proposed that TGF-β may recruit monocytes to differentiate into LCs in the periphery. In keeping with this hypothesis, mutant mice null for TGF-β lacked LCs. However, when irradiated TGF-$\beta^{+/+}$ mice were reconstituted with BM from TGF-$\beta^{-/-}$ mice, LCs were generated demonstrating that TGF-β produced by non-BM derived cells is required for the differentiation of LCs [35].

Direct evidence for a myeloid origin of DC came from studies demonstrating that the transplantation of mouse BM common myeloid progenitors (CMP) into irradiated recipients led to the reconstitution of the DC in the spleen, LNs and thymus [28, 30, 37]. Interestingly, the CMP was able to produce all cDC and pDC subsets found in the mouse spleen and thymus.

Lymphoid Origin of DC

Early studies by Vremec *et al.* (1992) [4] demonstrated that thymic cDC and subpopulations of cDC in mouse spleen and LNs expressed markers associated with lymphoid cells, including CD8α, CD4, CD2, BP1 and CD25. This finding suggested that some DC may have a lymphoid origin. Direct evidence was provided by studies showing that the transfer of an intrathymic lymphoid-restricted precursor – the CD4^{lo} precursor - into the thymus of irradiated recipients gave rise to both T cells and CD8$^+$ thymic cDC [38, 39]. This precursor was also shown to have the potential to produce both CD8$^+$ and CD8$^-$ splenic cDC when injected intravenously and to be devoid of myeloid potential [30, 40, 41]. In addition, the mouse BM common lymphoid progenitor (CLP) demonstrated the capacity to differentiate into DC both *in vitro* and *in vivo* [29, 30]. The CLP was shown to have the capacity to give rise to all splenic and thymic DC subsets, although a bias towards reconstituting the CD8$^+$ DC population was observed. Thus, evidence for a lymphoid origin of DC has now been firmly established.

The fact that both the CMP and CLP are capable of giving rise to all DC subtypes of the spleen and thymus not only demonstrates that DC can be of either myeloid or lymphoid origin, but also indicates that the phenotype of the DC is not a reflection of its lineage origin.

Early DC Precursors

The BM progenitors – CMP and CLP – are considered the earliest identified precursors of DC. However, it has been recently demonstrated that these populations of progenitors are in fact heterogenous. Both populations can be separated into two fractions based on their expression of Flt3 (receptor kinase fms-like tyrosine kinase 3) [42]. This study also demonstrated that the precursor activity for DC was only within the Flt3$^+$ fractions of both the CMP and CLP and these Flt3$^+$ precursors could give rise to all DC subsets in the spleen and thymus [42]. Therefore, this study revealed that the common feature of the earliest BM DC progenitors is the expression of Flt3 regardless of the lineage origin. This is an interesting finding given the well documented role for the ligand of Flt3 (Flt3L) in DC development. It has been shown that the addition of Flt3L in BM cell cultures promoted DC development [43-45] while the treatment of mice with Flt3L *in vivo* increased the number of both pDC and cDC in the blood and lymphoid tissues [44, 46, 47]. In addition, Flt3L$^{-/-}$ mice have reduced DC numbers [42, 48].

Immediate DC Precursors

The precursor cells that are at a developmental stage just before the formation of a phenotypically identifiable DC, is termed immediate DC precursor. They can be isolated from murine blood and have a capacity to differentiate into mature subsets of DC *in vitro*. Two populations of DC precursors have been described – CD11c^{int}CD45RA$^-$CD11b$^+$ and CD11c^{lo}CD45RA$^+$CD11b$^-$ [14]. The CD45RA$^-$ population are believed to be immature cDC, shown to acquire mature cDC morphology in the presence of TNF-α, stimulate T cells upon maturation and produce IL-12 in response to microbial stimuli. The CD45RA$^+$ population are believed to be the pDC [11-13, 49]. This population of cells matures into cDC-like cells only in the presence of stimuli such as CpG and GM-CSF, weakly stimulate T cells in this matured state and produce large quantities of type I interferon in response to CpG. This unique ability to produce type-1 interferons suggests that pDC are crucial for innate protection from virus infection. Thus, while this precursor DC population is able to differentiate into a cDC-like population of cells, their poor antigen presenting capabilities make them distinct from the mature cDC [50].

More recently, Naik *et al* [51] isolated a DC-restricted precursor population in the spleen capable of producing all splenic cDC but not pDC. This pre-cDC population is CD11c^{int}CD45RA^{lo}CD43^{int}Sirpα^{int}CD4$^-$CD8$^-$ and comprises 0.05% of splenocytes. When transferred into non-irradiated recipient mice, these pre-DC could generate all splenic cDC populations within 3-5 days representing the most immediate cDC precursor identified to date. Interestingly, this population was heterogeneous for CD24 – a marker co-expressed by the CD8$^+$ cDC but not the CD8$^-$ cDC. It was found that the CD24^{hi} pre-DC population generated the CD8$^+$ population while the CD24^{lo} population the CD8$^-$cDC. A CD24^{int} population gave rise to both subsets. Furthermore, this study showed that these pre-DC are not monocytes, because transfer of purified monocytes into a normal mouse did not generate significant number of cDC in the spleen. The monocytes could only differentiate into cDC under a GM-CSF mediated inflammatory condition. These intrasplenic pre-DC therefore represent the immediate cDC precursors in a steady-state mouse, while the monocytes can function as the immediate cDC precursors only under an inflammatory condition.

The identification of this intrasplenic pre-DC demonstrates that the spleen is capable of generating cDC *in situ*. The fact that the earlier BM progenitors – the CMP and CLP are able to generate both cDC and pDC, whilst the intrasplenic precursor is capable of generating cDC only, suggests that there must be a developmental branch downstream of the early BM progenitors that separates the two cDC and pDC lineages. This developmental branch is likely to occur in the bone marrow as both precursors of, and mature pDC are found in this site [52]. Whether a similar precursor population that gives rise to cDC *in situ* exists in other lymphoid tissues such as the lymph nodes, remains to be elucidated.

MOLECULAR REGULATION OF DC DEVELOPMENT

Given that all lymphoid-resident DC subsets can be produced from both myeloid and lymphoid precursors, the question still remains, how are the individual subsets delineated, and at what point in DC development is the commitment made to a subpopulation of DC? This question has been partially addressed with studies on transcription factors (TFs) in DC development. Growing evidence indicates that distinct DC subpopulations require different TFs during ontogeny. The first TF identified as being important in subset delineation was RelB. RelB is a member of the NF-kB/Rel family and was found to be expressed 6-fold higher in $CD8^-$ cDC compared to $CD8^+$ cDC in the spleen. Subsequently it was shown that this TF was required in the development of the splenic $CD8^-CD4^+$ DC but not the $CD8^+CD4^-$ and $CD8^-CD4^-$ DC [53-55]. Consistent with these observations, mice lacking the TRAF6 transcription factor, believed to act up stream of the same signalling cascade of RelB, also showed defects in $CD8^-CD4^+$ cDC development [56]. The zinc finger transcriptional regulator Ikaros has also been shown to play a role in DC development. While a dominant negative mutation in the Ikaros gene led to the ablation of all DC subsets [57], a null mutation in the Ikaros gene showed an absence of $CD8^-$ cDC and pDC and residual $CD8^+$ cDC development [58-60]. These results indicated that Ikaros was required for the development of most cDC and pDC. This finding is not surprising, because it has been shown that Ikraros is required for the normal development of the early hemopoietic progenitors [61].

More recently, three interferon regulatory factors (IRF), namely IRF-2, IRF-4 and IRF-8 have been identified in playing important roles in the development of different DC populations. Mice lacking a functional IRF-2 or IRF-4 gene show defects in the $CD8^-CD4^+$ cDC and pDC subsets in the spleen [62-64]. In contrast, IRF-8 (also known as ICSBP) has been found to play crucial roles in the development and function of $CD8^+CD4^-$ cDC, pDC and LC [65, 66]. Another TF, Id2, a member of the helix-loop-helix TF family, has been shown to be required for the development of $CD8^+CD4^-$ cDC and LC [67]. In addition, the transcription factor SpiB has been shown to play a crucial role in pDC development [68].

A recent study revealed the requirement of distinct molecular signalling pathways for the Flt3L-dependent steady-state DC development and the inflammation induced GM-CSF-dependent DC differentiation [69]. This study demonstrated that the transcription factor STAT3 was indispensable in Flt3L-regulated DC development. Ablation of STAT3 in hemopoietic-derived cells led to a profound deficiency in the DC compartment and abolished the effects of Flt3L on DC generation. This study also showed that, in contrast to the FLt3L-dependent DC differentiation pathway, activation of STAT3 was not required for GM-CSF-dependent DC differentiation *in vitro*.

FUNCTIONAL DIVERSITIES OF DC SUBSETS

In general, all cDC have the ability to take up, process and present antigen to naïve T cells, representing the most potent of the antigen presenting cells (APC) [1]. Presentation of antigen to T cells can achieve two different outcomes. In the steady state, DC continuously process and present endogenous self-antigen to T cells. T cells that are highly reactive to self-antigens are deleted either within the thymus through a process called "negative selection" [70], or in the periphery in a process called peripheral tolerance. The function of DC in the steady state is overcome when DC become activated and mature after engulfing a foreign antigen, or in response to inflammatory signals such as interleukin-1 (IL-1) and tumour necrosis factor-α (TNFα) that are released at the site of an infection. Activated DC are then able to prime and activate naïve T cells, thereby initiating an adaptive immune response.

Although all DC function as potent APCs, the heterogeneity of the DC population strongly suggests a role for individual DC subsets in different immune functions. Certain DC subsets have been suggested to play a role in maintaining tolerance to self-antigens [71-73], while other DC subtypes were shown to be responsible for priming the innate and adaptive immune responses [74, 75]. Although there is increasing evidence to support the functional specialisation of individual DC subset, the system appears to be more plastic and complex.

DC are able to polarize CD4^{+} T cells to become either a Th1 or a Th2 cell. Th1 responses are characterized by high levels of interferon-γ (IFN-γ) production by CD4^{+} T cells [76, 77] and are involved in eliminating intracellular pathogens directly and via macrophage activation [78]. In contrast, Th2 responses are characterized by the production of IL-4 and IL-10 by CD4^{+} T cells [76, 77] and are responsible for eliminating extracellular pathogens via activation of cognate B cells [78]. The differential polarization of T cells requires a different array of cytokines produced by activated cDC. IL-12 production by cDC causes Th1 polarization, whereas the presence of IL-10 suppresses IL-12 production by cDC thereby favouring Th2 polarization. It has been shown that *in vitro* mouse splenic CD4^{-}8^{+} cDC produce large quantities of IL-12 after activation compared to CD4^{-}8^{-} cDC and CD4^{+}8^{-} cDC [79]. In addition, *in vivo* studies have shown that CD8^{+} cDC can prime Th1 responses, whereas CD8^{-} cDC prime Th2 responses [75, 80, 81]. Although a clear functional specification of different cDC subsets was observed in these experimental systems, other extrinsic factors such as the location of DC, the microenvironment, the types of stimuli or pathogens they encounter and the kinetics of DC activation can also influence the functional outcomes of a given DC subset [82-84].

A role for DC in maintaining peripheral tolerance to self-antigen in the steady state has been proposed. Evidence suggests that this process is carried out mainly by the CD4^{-}8^{+} cDC subset. The CD4^{-}8^{+} cDC have been shown to be able to induce CD4^{+} T cell death *in vitro* [73] and to present self-antigen by endocytosing apoptotic cells *in vivo* in the steady state and to tolerise self- reactive T cells in the periphery [85]. In order for tolerance to occur successfully, the DC must present self, endogenous antigen on both MHC Class-I and MHC Class-II molecules to CD8^{+} and CD4^{+} T cells respectively. Exogenous antigen is usually presented on MHC Class-II molecules to CD4^{+} T cells and endogenous antigen on MHC Class-I to CD8^{+} T cells. The process by which DC can present exogenous antigen on MHC Class-I molecules to CD8^{+} T cells is known as cross-presentation [86]. The CD4^{-}8^{+} cDCs have demonstrated a unique ability to cross present exogenous antigen (e.g. endocytosed apoptotic cells) for CD8$^+$ T cell responses *in vivo* [87]. This function has been suggested to be responsible for the ability of CD4$^-$8$^+$ cDC to induce peripheral tolerance to tissue associated antigens [86].

The cross-presentation capabilities of the CD4$^-$8$^+$ cDC have also been shown to be important in viral immunity. In a viral skin infection model, it was demonstrated that LCs carried the virus from the sight of infection to the draining LNs where the viral antigens were then picked up by the LN resident CD4$^-$8$^+$ cDC population and cross-presented to CD8$^+$ T cells to initiate a cytotoxic T cell response [88]. These findings have also been demonstrated in other viral infection models where different administration routes have been used [89].

Recently a unique function for the splenic CD8$^-$CD4$^+$ cDC has been described. It was shown that in addition to their ability to present antigens to naïve T cells, the splenic CD8$^-$CD4$^+$ cDC are also the major producers of inflammatory chemokines Mip1-α, Mip1-β and Rantes in response to TLR stimulation *in vitro* [90].

The evidence for functional specialization of DC subsets described above was mainly from the studies of mouse splenic cDC subsets. The thymic DC, although sharing many common features of splenic DC, have been suggested to have different functional capacities. Thymic cDC play important roles in the induction of central immune tolerance via a process known as "negative selection" of developing thymocytes [91-93]. Thymic cDC can present self-antigen to developing thymocytes and the thymocytes with a high affinity TCR to self-antigen will be deleted within the thymus. Thymic cDC also have the ability to induce T regulatory cells, which may be partly responsible for the maintenance of peripheral tolerance [94]. Thymic cDC contain at least two different subsets: one developed *in situ* and representing ~70% of thymic cDC [5], the other minor subset may originate from periphery and migrate to the thymus [95]. Two phenotypically different mouse thymic cDC subsets have been described recently, namely the CD8$^+$Sirpα^- (~70%) and the CD8$^{-/lo}$Sirpα^+ cDC (~30%) [5]. Whether these two cDC subsets correspond to the *in situ* developed and the migratory DC subset respectively is currently not clear. Further functional studies of these thymic cDC subsets will be an interesting subject in the field of DC biology and will provide new insights into the cellular mechanisms in central tolerance induction.

GENERATION OF STEADY-STATE DC SUBSET EQUIVALENTS *IN VITRO*

The studies of functional capacities of DC subsets were mainly performed using *ex-vivo* isolated DC populations. However, the diminutive numbers of DC and difficulty in their isolation has often precluded more detailed analysis of their biology. This issue has been addressed by developing culture methods that generate DC in culture in the presence of various cytokines. The first such method developed utilized monocytes as a precursor of DC in the presence of GM-CSF and IL-4 [96]. The monocyte-derived DC (moDC) produced in this culture system have been extensively studied, and used in a number of DC-based vaccination trials in humans. MoDC are a homogeneous population of cDC that express CD11c, high levels of MHC Class-I and II molecules, and are stimulatory in a mixed leukocyte reaction (MLR).

While this system has proved useful in studying some aspects of DC biology, and to date represents the easiest method of generating large numbers of DC in culture, the resulting DC do not represent the heterogeneous DC subpopulations identified *in vivo*. Furthermore, it was demonstrated recently in mice, that the monocyte to DC conversion does not occur on a continual basis in the steady state *in vivo* but mainly during an inflammatory response when levels of serum GM-CSF are high [51].

A more recent culture system for generating DC involves culturing bone marrow cells in the presence of Flt3 ligand. This system yields both cDC and pDC in large numbers [45, 97]. Given the important role of Flt3L in DC development *in vivo*, this system represents a more physiological approach to generating steady state DC *in vitro*. Moreover, the cDC produced in this culture are heterogeneous in phenotype and can be clearly segregated into two populations: a $CD24^{hi}CD11b^{lo}Sirp\alpha^-$ and $CD24^{lo}CD11b^{hi}Sirp\alpha^+$ population [97]. These two populations were shown to be the phenotypically and functional equivalents of steady-state splenic $CD8^+$ and $CD8^-$ cDC subsets based on their surface marker expression, transcription factor expression and dependence for development, ability to cross-present cellular antigen to CD8 T cells, cytokine production and expression of chemokine and toll-like receptors. pDC were also produced in this culture system and demonstrated a high capacity for IFN-α production in response to TLR stimulation. This culture system yields up to 25×10^6 $CD8^+$ cDC from the BM of one mouse compared to 1×10^6 $CD8^+$ cDC from one mouse spleen. Such a system allows more detailed studies of DC development and function when large number of cells are required.

CONCLUDING REMARKS

As summarized in this review, DC represent a cell system consisting of several phenotypic and functional distinct subsets. Although the ultimate origin of all cDC and pDC is the BM hemopoietic stem cells (HSC), DC can develop through either the myeloid or lymphoid pathways from a Flt3 expressing progenitor (Figure 2). The transcription factors, including Ikaros, RelB, IRF-2, IRF-4, IRF-8, Id2 and SpiB play important roles in controlling the development of different DC populations. Based on the differential expression of several surface molecules, DC in mouse lymphoid organs can be segregated into several subsets that differ in the capacities of antigen presentation, T cell activation, cytokine and chemokine production, which leads to different outcomes of immune responses, i.e. immunity or tolerance. The intrinsic property of each DC subset may dictate their functional specialty, extrinsic factors should also be considered as additional determinants of DC function. A better understanding of the detailed processes governing the development and function of these DC subsets is essential for future application of DC in immune modulation, vaccine design and immune therapies.

REFERENCES

[1]   Steinman RM. The dendritic cell system and its role in immunogenicity. *Annu. Rev. Immunol.* 1991;9:271-96.

[2]   Shortman K, Liu YJ. Mouse and human dendritic cell subtypes. *Nat. Rev. Immunol.* 2002;2(3):151-61.

[3]   O'Keeffe M, Hochrein H, Vremec D, Caminschi I, Miller JL, Anders EM, et al. Mouse plasmacytoid cells: long-lived cells, heterogeneous in surface phenotype and function, that differentiate into CD8(+) dendritic cells only after microbial stimulus. *J. Exp. Med.* 2002;196(10):1307-19.

[4]   Vremec D, Zorbas M, Scollay R, Saunders DJ, Ardavin CF, Wu L, et al. The surface phenotype of dendritic cells purified from mouse thymus and spleen: investigation of the CD8 expression by a subpopulation of dendritic cells. *J. Exp. Med.* 1992;176(1):47-58.

[5]   Lahoud MH, Proietto AI, Gartlan KH, Kitsoulis S, Curtis J, Wettenhall J, et al. Signal regulatory protein molecules are differentially expressed by CD8- dendritic cells. *J. Immunol.* 2006;177(1):372-82.

[6]   Vremec D, Pooley J, Hochrein H, Wu L, Shortman K. CD4 and CD8 expression by dendritic cell subtypes in mouse thymus and spleen. *J. Immunol.* 2000;164(6):2978-86.

[7]   Henri S, Vremec D, Kamath A, Waithman J, Williams S, Benoist C, et al. The dendritic cell populations of mouse lymph nodes. *J. Immunol.* 2001;167(2):741-8.

[8]   Anjuere F, Martin P, Ferrero I, Fraga ML, del Hoyo GM, Wright N, et al. Definition of dendritic cell subpopulations present in the spleen, Peyer's patches, lymph nodes, and skin of the mouse. *Blood* 1999;93(2):590-8.

[9]   Kamath AT, Pooley J, O'Keeffe MA, Vremec D, Zhan Y, Lew AM, et al. The development, maturation, and turnover rate of mouse spleen dendritic cell populations. *J. Immunol.* 2000;165(12):6762-70.

[10]   Iwasaki A, Kelsall BL. Localization of distinct Peyer's patch dendritic cell subsets and their recruitment by chemokines macrophage inflammatory protein (MIP)-3alpha, MIP-3beta, and secondary lymphoid organ chemokine. *J. Exp. Med.* 2000;191(8):1381-94.

[11]   Asselin-Paturel C, Boonstra A, Dalod M, Durand I, Yessaad N, Dezutter-Dambuyant C, et al. Mouse type I IFN-producing cells are immature APCs with plasmacytoid morphology. *Nat. Immunol.* 2001;2(12):1144-50.

[12]   Siegal FP, Kadowaki N, Shodell M, Fitzgerald-Bocarsly PA, Shah K, Ho S, et al. The nature of the principal type 1 interferon-producing cells in human blood. *Science* 1999;284(5421):1835-7.

[13]   Nakano H, Yanagita M, Gunn MD. CD11c(+)B220(+)Gr-1(+) cells in mouse lymph nodes and spleen display characteristics of plasmacytoid dendritic cells. *J. Exp. Med.* 2001;194(8):1171-8.

[14]   O'Keeffe M, Hochrein H, Vremec D, Scott B, Hertzog P, Tatarczuch L, et al. Dendritic cell precursor populations of mouse blood: identification of the murine homologues of human blood plasmacytoid pre-DC2 and CD11c+ DC1 precursors. *Blood* 2003;101(4):1453-9.

[15]   Bjorck P. Isolation and characterization of plasmacytoid dendritic cells from Flt3 ligand and granulocyte-macrophage colony-stimulating factor-treated mice. *Blood* 2001;98(13):3520-6.

[16]   Dakic A, Shao QX, D'Amico A, O'Keeffe M, Chen WF, Shortman K, et al. Development of the dendritic cell system during mouse ontogeny. *J. Immunol.* 2004;172(2):1018-27.

[17] Dakic A, Wu L. Hemopoietic precursors and development of dendritic cell populations. *Leuk. Lymphoma* 2003;44(9):1469-75.

[18] Muthukkumar S, Goldstein J, Stein KE. The ability of B cells and dendritic cells to present antigen increases during ontogeny. *J. Immunol.* 2000;165(9):4803-13.

[19] Ridge JP, Fuchs EJ, Matzinger P. Neonatal tolerance revisited: turning on newborn T cells with dendritic cells. *Science* 1996;271(5256):1723-6.

[20] Sun CM, Fiette L, Tanguy M, Leclerc C, Lo-Man R. Ontogeny and innate properties of neonatal dendritic cells. *Blood* 2003;102(2):585-91.

[21] Vollstedt S, Franchini M, Hefti HP, Odermatt B, O'Keeffe M, Alber G, et al. Flt3 ligand-treated neonatal mice have increased innate immunity against intracellular pathogens and efficiently control virus infections. *J. Exp. Med.* 2003;197(5):575-84.

[22] Jones RJ, Collector MI, Barber JP, Vala MS, Fackler MJ, May WS, et al. Characterization of mouse lymphohematopoietic stem cells lacking spleen colony-forming activity. *Blood* 1996;88(2):487-91.

[23] Eaves C, Miller C, Cashman J, Conneally E, Petzer A, Zandstra P, et al. Hematopoietic stem cells: inferences from in vivo assays. *Stem Cells* 1997;15 Suppl 1:1-5.

[24] Lemischka I. Stem cell dogmas in the genomics era. *Rev. Clin. Exp. Hematol.* 2001;5(1):15-25.

[25] Weissman IL. Stem cells: units of development, units of regeneration, and units in evolution. *Cell* 2000;100(1):157-68.

[26] Akashi K, Traver D, Miyamoto T, Weissman IL. A clonogenic common myeloid progenitor that gives rise to all myeloid lineages. *Nature* 2000;404(6774):193-7.

[27] Kondo M, Weissman IL, Akashi K. Identification of clonogenic common lymphoid progenitors in mouse bone marrow. *Cell* 1997;91(5):661-72.

[28] Manz MG, Traver D, Akashi K, Merad M, Miyamoto T, Engleman EG, et al. Dendritic cell development from common myeloid progenitors. *Ann. NY Acad. Sci.* 2001;938:167-73; discussion 173-4.

[29] Manz MG, Traver D, Miyamoto T, Weissman IL, Akashi K. Dendritic cell potentials of early lymphoid and myeloid progenitors. *Blood* 2001;97(11):3333-41.

[30] Wu L, D'Amico A, Hochrein H, O'Keeffe M, Shortman K, Lucas K. Development of thymic and splenic dendritic cell populations from different hemopoietic precursors. *Blood* 2001;98(12):3376-82.

[31] Inaba K, Inaba M, Deguchi M, Hagi K, Yasumizu R, Ikehara S, et al. Granulocytes, macrophages, and dendritic cells arise from a common major histocompatibility complex class II-negative progenitor in mouse bone marrow. *Proc. Natl. Acad. Sci. USA* 1993;90(7):3038-42.

[32] Reid CD, Stackpoole A, Meager A, Tikerpae J. Interactions of tumor necrosis factor with granulocyte-macrophage colony-stimulating factor and other cytokines in the regulation of dendritic cell growth in vitro from early bipotent CD34+ progenitors in human bone marrow. *J. Immunol.* 1992;149(8):2681-8.

[33] Szabolcs P, Avigan D, Gezelter S, Ciocon DH, Moore MA, Steinman RM, et al. Dendritic cells and macrophages can mature independently from a human bone marrow-derived, post-colony-forming unit intermediate. *Blood* 1996;87(11):4520-30.

[34] Caux C, Vanbervliet B, Massacrier C, Dezutter-Dambuyant C, de Saint-Vis B, Jacquet C, et al. CD34+ hematopoietic progenitors from human cord blood differentiate along

two independent dendritic cell pathways in response to GM-CSF+TNF alpha. *J. Exp. Med.* 1996;184(2):695-706.

[35]   Geissmann F, Prost C, Monnet JP, Dy M, Brousse N, Hermine O. Transforming growth factor beta1, in the presence of granulocyte/macrophage colony-stimulating factor and interleukin 4, induces differentiation of human peripheral blood monocytes into dendritic Langerhans cells. *J. Exp. Med.* 1998;187(6):961-6.

[36]   Randolph GJ, Inaba K, Robbiani DF, Steinman RM, Muller WA. Differentiation of phagocytic monocytes into lymph node dendritic cells in vivo. *Immunity* 1999;11(6):753-61.

[37]   Traver D, Akashi K, Manz M, Merad M, Miyamoto T, Engleman EG, et al. Development of CD8alpha-positive dendritic cells from a common myeloid progenitor. *Science* 2000;290(5499):2152-4.

[38]   Wu L, Vremec D, Ardavin C, Winkel K, Suss G, Georgiou H, et al. Mouse thymus dendritic cells: kinetics of development and changes in surface markers during maturation. *Eur J Immunol* 1995;25(2):418-25.

[39]   Ardavin C, Wu L, Li CL, Shortman K. Thymic dendritic cells and T cells develop simultaneously in the thymus from a common precursor population. *Nature* 1993;362(6422):761-3.

[40]   Martin P, del Hoyo GM, Anjuere F, Ruiz SR, Arias CF, Marin AR, et al. Concept of lymphoid versus myeloid dendritic cell lineages revisited: both CD8alpha(-) and CD8alpha(+) dendritic cells are generated from CD4(low) lymphoid-committed precursors. *Blood* 2000;96(7):2511-9.

[41]   Izon D, Rudd K, DeMuth W, Pear WS, Clendenin C, Lindsley RC, et al. A common pathway for dendritic cell and early B cell development. *J. Immunol.* 2001;167(3):1387-92.

[42]   D'Amico A, Wu L. The early progenitors of mouse dendritic cells and plasmacytoid predendritic cells are within the bone marrow hemopoietic precursors expressing Flt3. *J. Exp. Med.* 2003;198(2):293-303.

[43]   Saunders D, Lucas K, Ismaili J, Wu L, Maraskovsky E, Dunn A, et al. Dendritic cell development in culture from thymic precursor cells in the absence of granulocyte/macrophage colony-stimulating factor. *J. Exp. Med.* 1996;184(6):2185-96.

[44]   O'Keeffe M, Hochrein H, Vremec D, Pooley J, Evans R, Woulfe S, et al. Effects of administration of progenipoietin 1, Flt-3 ligand, granulocyte colony-stimulating factor, and pegylated granulocyte-macrophage colony-stimulating factor on dendritic cell subsets in mice. *Blood* 2002;99(6):2122-30.

[45]   Brasel K, De Smedt T, Smith JL, Maliszewski CR. Generation of murine dendritic cells from flt3-ligand-supplemented bone marrow cultures. *Blood* 2000;96(9):3029-39.

[46]   Drakes ML, Lu L, Subbotin VM, Thomson AW. In vivo administration of flt3 ligand markedly stimulates generation of dendritic cell progenitors from mouse liver. *J. Immunol.* 1997;159(9):4268-78.

[47]   Maraskovsky E, Brasel K, Teepe M, Roux ER, Lyman SD, Shortman K, et al. Dramatic increase in the numbers of functionally mature dendritic cells in Flt3 ligand-treated mice: multiple dendritic cell subpopulations identified. *J. Exp. Med.* 1996; 184(5):1953-62.

[48] McKenna HJ, Stocking KL, Miller RE, Brasel K, De Smedt T, Maraskovsky E, et al. Mice lacking flt3 ligand have deficient hematopoiesis affecting hematopoietic progenitor cells, dendritic cells, and natural killer cells. *Blood* 2000;95(11):3489-97.

[49] Martin P, Del Hoyo GM, Anjuere F, Arias CF, Vargas HH, Fernandez LA, et al. Characterization of a new subpopulation of mouse CD8alpha+ B220+ dendritic cells endowed with type 1 interferon production capacity and tolerogenic potential. *Blood* 2002;100(2):383-90.

[50] Krug A, Veeraswamy R, Pekosz A, Kanagawa O, Unanue ER, Colonna M, et al. Interferon-producing cells fail to induce proliferation of naive T cells but can promote expansion and T helper 1 differentiation of antigen-experienced unpolarized T cells. *J. Exp. Med.* 2003;197(7):899-906.

[51] Naik SH, Metcalf D, van Nieuwenhuijze A, Wicks I, Wu L, O'Keeffe M, et al. Intrasplenic steady-state dendritic cell precursors that are distinct from monocytes. *Nat. Immunol.* 2006;7(6):663-71.

[52] Naik SH, Corcoran LM, Wu L. Development of murine plasmacytoid dendritic cell subsets. *Immunol. Cell Biol.* 2005;83(5):563-70.

[53] Wu L, D'Amico A, Winkel KD, Suter M, Lo D, Shortman K. RelB is essential for the development of myeloid-related CD8alpha- dendritic cells but not of lymphoid-related CD8alpha+ dendritic cells. *Immunity* 1998;9(6):839-47.

[54] Weih F, Carrasco D, Durham SK, Barton DS, Rizzo CA, Ryseck RP, et al. Multiorgan inflammation and hematopoietic abnormalities in mice with a targeted disruption of RelB, a member of the NF-kappa B/Rel family. *Cell* 1995;80(2):331-40.

[55] Burkly L, Hession C, Ogata L, Reilly C, Marconi LA, Olson D, et al. Expression of relB is required for the development of thymic medulla and dendritic cells. *Nature* 1995;373(6514):531-6.

[56] Kobayashi T, Walsh PT, Walsh MC, Speirs KM, Chiffoleau E, King CG, et al. TRAF6 is a critical factor for dendritic cell maturation and development. *Immunity* 2003;19(3):353-63.

[57] Wu L, Nichogiannopoulou A, Shortman K, Georgopoulos K. Cell-autonomous defects in dendritic cell populations of Ikaros mutant mice point to a developmental relationship with the lymphoid lineage. *Immunity* 1997;7(4):483-92.

[58] Georgopoulos K, Winandy S, Avitahl N. The role of the Ikaros gene in lymphocyte development and homeostasis. *Annu Rev Immunol* 1997;15:155-76.

[59] Georgopoulos K, Bigby M, Wang JH, Molnar A, Wu P, Winandy S, et al. The Ikaros gene is required for the development of all lymphoid lineages. *Cell* 1994;79(1):143-56.

[60] Allman D, Dalod M, Asselin-Paturel C, Delale T, Robbins SH, Trinchieri G, et al. Ikaros is required for plasmacytoid dendritic cell differentiation. *Blood* 2006;

[61] Shortman K, Wu L. Early T lymphocyte progenitors. *Annu. Rev. Immunol.* 1996;14:29-47.

[62] Suzuki S, Honma K, Matsuyama T, Suzuki K, Toriyama K, Akitoyo I, et al. Critical roles of interferon regulatory factor 4 in CD11bhighCD8alpha- dendritic cell development. *Proc. Natl. Acad. Sci. USA* 2004;101(24):8981-6.

[63] Ichikawa E, Hida S, Omatsu Y, Shimoyama S, Takahara K, Miyagawa S, et al. Defective development of splenic and epidermal CD4+ dendritic cells in mice deficient for IFN regulatory factor-2. *Proc. Natl. Acad. Sci. USA* 2004;101(11):3909-14.

[64] Honda K, Mizutani T, Taniguchi T. Negative regulation of IFN-alpha/beta signaling by IFN regulatory factor 2 for homeostatic development of dendritic cells. *Proc. Natl. Acad. Sci. USA* 2004;101(8):2416-21.

[65] Schiavoni G, Mattei F, Borghi P, Sestili P, Venditti M, Morse HC, 3rd, et al. ICSBP is critically involved in the normal development and trafficking of Langerhans cells and dermal dendritic cells. *Blood* 2004;103(6):2221-8.

[66] Schiavoni G, Mattei F, Sestili P, Borghi P, Venditti M, Morse HC, 3rd, et al. ICSBP is essential for the development of mouse type I interferon-producing cells and for the generation and activation of CD8alpha(+) dendritic cells. *J. Exp. Med.* 2002;196(11):1415-25.

[67] Hacker C, Kirsch RD, Ju XS, Hieronymus T, Gust TC, Kuhl C, et al. Transcriptional profiling identifies Id2 function in dendritic cell development. *Nat. Immunol.* 2003;4(4):380-6.

[68] Schotte R, Nagasawa M, Weijer K, Spits H, Blom B. The ETS transcription factor Spi-B is required for human plasmacytoid dendritic cell development. *J. Exp. Med.* 2004;200(11):1503-9.

[69] Laouar Y, Welte T, Fu XY, Flavell RA. STAT3 is required for Flt3L-dependent dendritic cell differentiation. *Immunity* 2003;19(6):903-12.

[70] Brocker T, Riedinger M, Karjalainen K. Targeted expression of major histocompatibility complex (MHC) class II molecules demonstrates that dendritic cells can induce negative but not positive selection of thymocytes in vivo. *J. Exp. Med.* 1997;185(3):541-50.

[71] Kronin V, Winkel K, Suss G, Kelso A, Heath W, Kirberg J, et al. A subclass of dendritic cells regulates the response of naive CD8 T cells by limiting their IL-2 production. *J. Immunol.* 1996;157(9):3819-27.

[72] Steinman RM, Hawiger D, Liu K, Bonifaz L, Bonnyay D, Mahnke K, et al. Dendritic cell function in vivo during the steady state: a role in peripheral tolerance. *Ann. NY Acad. Sci.* 2003;987:15-25.

[73] Suss G, Shortman K. A subclass of dendritic cells kills CD4 T cells via Fas/Fas-ligand-induced apoptosis. *J Exp Med* 1996;183(4):1789-96.

[74] Maldonado-Lopez R, De Smedt T, Michel P, Godfroid J, Pajak B, Heirman C, et al. CD8alpha+ and CD8alpha- subclasses of dendritic cells direct the development of distinct T helper cells in vivo. *J. Exp. Med.* 1999;189(3):587-92.

[75] Pulendran B, Smith JL, Caspary G, Brasel K, Pettit D, Maraskovsky E, et al. Distinct dendritic cell subsets differentially regulate the class of immune response in vivo. *Proc Natl Acad Sci U S A* 1999;96(3):1036-41.

[76] Seder RA. Acquisition of lymphokine-producing phenotype by CD4+ T cells. *J. Allergy Clin. Immunol.* 1994;94(6 Pt 2):1195-202.

[77] Mosmann TR, Coffman RL. TH1 and TH2 cells: different patterns of lymphokine secretion lead to different functional properties. *Annu. Rev. Immunol.* 1989;7:145-73.

[78] Constant SL, Bottomly K. Induction of Th1 and Th2 CD4+ T cell responses: the alternative approaches. *Annu. Rev. Immunol.* 1997;15:297-322.

[79] Hochrein H, Shortman K, Vremec D, Scott B, Hertzog P, O'Keeffe M. Differential production of IL-12, IFN-alpha, and IFN-gamma by mouse dendritic cell subsets. *J. Immunol* .2001;166(9):5448-55.

[80] Moser M, Murphy KM. Dendritic cell regulation of TH1-TH2 development. *Nat Immunol* 2000;1(3):199-205.

[81] Maldonado-Lopez R, Moser M. Dendritic cell subsets and the regulation of Th1/Th2 responses. *Semin. Immunol.* 2001;13(5):275-82.

[82] Pulendran B, Kumar P, Cutler CW, Mohamadzadeh M, Van Dyke T, Banchereau J. Lipopolysaccharides from distinct pathogens induce different classes of immune responses in vivo. *J. Immunol.* 2001;167(9):5067-76.

[83] Manickasingham SP, Edwards AD, Schulz O, Reis e Sousa C. The ability of murine dendritic cell subsets to direct T helper cell differentiation is dependent on microbial signals. *Eur. J. Immunol.* 2003;33(1):101-7.

[84] Langenkamp A, Messi M, Lanzavecchia A, Sallusto F. Kinetics of dendritic cell activation: impact on priming of TH1, TH2 and nonpolarized T cells. *Nat. Immunol.* 2000;1(4):311-6.

[85] Iyoda T, Shimoyama S, Liu K, Omatsu Y, Akiyama Y, Maeda Y, et al. The CD8+ dendritic cell subset selectively endocytoses dying cells in culture and in vivo. *J. Exp. Med.* 2002;195(10):1289-302.

[86] Belz GT, Behrens GM, Smith CM, Miller JF, Jones C, Lejon K, et al. The CD8alpha(+) dendritic cell is responsible for inducing peripheral self-tolerance to tissue-associated antigens. *J. Exp. Med.* 2002; 196(8):1099-104.

[87] den Haan JM, Lehar SM, Bevan MJ. CD8(+) but not CD8(-) dendritic cells cross-prime cytotoxic T cells in vivo. *J. Exp. Med.* 2000;192(12):1685-96.

[88] Allan RS, Waithman J, Bedoui S, Jones CM, Villadangos JA, Zhan Y, et al. Migratory dendritic cells transfer antigen to a lymph node-resident dendritic cell population for efficient CTL priming. *Immunity* 2006;25(1):153-62.

[89] Belz GT, Smith CM, Eichner D, Shortman K, Karupiah G, Carbone FR, et al. Cutting Edge: Conventional CD8alpha(+) Dendritic Cells Are Generally Involved in Priming CTL Immunity to Viruses. *J. Immunol.* 2004;172(4):1996-2000.

[90] Proietto AI, O'Keeffe M, Gartlan K, Wright MD, Shortman K, Wu L, et al. Differential production of inflammatory chemokines by murine dendritic cell subsets. *Immunobiology* 2004;209(1-2):163-72.

[91] Gao EK, Lo D, Sprent J. Strong T cell tolerance in parent----F1 bone marrow chimeras prepared with supralethal irradiation. Evidence for clonal deletion and anergy. *J. Exp. Med.* 1990;171(4):1101-21.

[92] Gallegos AM, Bevan MJ. Central tolerance to tissue-specific antigens mediated by direct and indirect antigen presentation. *J. Exp. Med.* 2004;200(8):1039-49.

[93] Bonasio R, Scimone ML, Schaerli P, Grabie N, Lichtman AH, von Andrian UH. Clonal deletion of thymocytes by circulating dendritic cells homing to the thymus. *Nat. Immunol.* 2006;7(10):1092-100.

[94] Watanabe N, Wang YH, Lee HK, Ito T, Cao W, Liu YJ. Hassall's corpuscles instruct dendritic cells to induce CD4+CD25+ regulatory T cells in human thymus. *Nature* 2005;436(7054):1181-5.

[95] Goldschneider I, Cone RE. A central role for peripheral dendritic cells in the induction of acquired thymic tolerance. *Trends Immunol* 2003;24(2):77-81.

[96] Sallusto F, Lanzavecchia A. Efficient presentation of soluble antigen by cultured human dendritic cells is maintained by granulocyte/macrophage colony-stimulating factor plus

interleukin 4 and downregulated by tumor necrosis factor alpha. *J. Exp. Med.* 1994;179(4):1109-18.

[97] Naik SH, Proietto AI, Wilson NS, Dakic A, Schnorrer P, Fuchsberger M, et al. Cutting edge: generation of splenic CD8+ and CD8- dendritic cell equivalents in Fms-like tyrosine kinase 3 ligand bone marrow cultures. *J. Immunol.* 2005;174(11):6592-7.

In: Stem Cell Research Progress
Editor: Prasad S. Koka, pp. 137-157

ISBN: 978-1-60456-308-5
© 2008 Nova Science Publishers, Inc.

Chapter 9

ADVANCES TOWARDS THE DEVELOPMENT OF A STEM CELL THERAPY FOR HIRSCHSPRUNG'S DISEASE

Daniel B. Hawcutt[1,2], Richard M. Lindley[1,2], Simon E. Kenny[2] and David H. Edgar[1] *

[1]School of Biomedical Sciences, The University of Liverpool, Sherrington Buildings, Liverpool L69 3GE, UK
[2]Institute of Child Health, Royal Liverpool University Children's Hospital, Alder Hey, Liverpool L12 2AP

ABSTRACT

There has been considerable progress towards the development of a stem cell-based therapy for Hirschsprung's disease (HSCR). This review article discusses the mechanisms regulating enteric nervous system (ENS) stem cell development and the molecular pathogenesis of HSCR in order to elucidate the feasibility of any stem cell therapy. Such a therapy will require a source of stem cells, techniques to isolate and amplify them, and effective methods for the transplantation of the stem cells or their progeny. To date, embryonic and postnatal ENS stem cells have been cultured as neurospheres prior to transplantion into mammalian bowel. However, while there are still many obstacles to overcome, recent observations have demonstrated functional changes in the gut after stem cell transplantation, indicating that a stem cell therapy for HSCR will indeed be possible.

Keywords: Neural crest, neuronal stem cells, transplantation, neurosphere, Hirschsprung's disease

---

* Correspondence: David H Edgar, Ph.D. School of Biomedical Sciences, The University of Liverpool, Sherrington Buildings, Liverpool L69 3GE, UK. Phone: +441517945508; Fax +441517945508. E-mail: dhedgar@liv.ac.uk

INTRODUCTION

The Need for a Stem Cell Therapy for Hirschsprung's Disease

Hirschsprung's disease (HSCR) is the most common developmental disorder of the neural crest in humans, with an incidence of approximately 1 in 5000 live births [1]. In HSCR, a variable length of the terminal gut lacks ganglion cells of the enteric nervous system (ENS), due to failure of ENS stem cells to fully colonize the distal portion of the gut during embryonic development. This lack of neurons adversely affects normal peristalsis in the distal gut, resulting in a state of chronic contraction that can cause a life-threatening bowel obstruction.

HSCR may occur in isolation or as part of a broader clinical picture, including Down's, Waardenburg-Shah or Haddad syndromes. Whether syndromic or an isolated case, surgical treatment remains essentially unchanged since its first description, involving resection of the affected gut and 'pull-through' of normally innervated intestine to the anus [2]. However, the long term outcome of surgery is often unsatisfactory, mainly due to complications including incontinence [3, 4]. Clearly, new approaches to treatment are required in order to improve the quality of life of patients. One possibility may be to use stem cells to colonize those parts of the gut lacking an ENS. An understanding of the mechanisms of enteric nervous system development and the molecular pathogenesis of HSCR are essential to the development of successful stem cell based therapies. This review will therefore briefly cover the ontogeny of ENS stem cells before detailing various approaches to apply stem cell therapy for HSCR. In addition, the areas where further research is needed in order to achieve the goal of a successful stem cell therapy will be highlighted.

The Enteric Nervous System and Its Development

The ENS is part of the peripheral nervous system and extends the entire length of the gastrointestinal tract in which it regulates both gut motility and fluid and electrolyte balance [5]. As the ENS is phylogenetically well conserved across vertebrate classes [6, 7], there are several animal model systems relevant to the study of HSCR.

Neurons and glia of the ENS are clustered in ganglia located in the myenteric (Auerbach's) plexus and the submucosal (Meissner's) plexus of the gut wall. These enteric ganglia contain many types of neuron producing a wide variety of excitatory and inhibitory neurotransmitters that regulate smooth muscle contraction, alter secretions into the lumen of the gut, and receive or transmit information to other parts of the nervous system. Although ganglia receive input from other parts of the nervous system, the ENS is capable of full autonomy. There are estimated to be over 100 million neurons in the human gut [8], together with enteric glia that outnumber the neurons. The glia function as a store for neurotransmitter precursors, uptake and degrade neurotransmitters, express receptors for the neurotransmitters, and maintain the integrity of the mucosal barrier of the gut [9].

One of the challenges of developmental biology is, therefore, to determine how stem cells of the ENS give rise to such anatomical complexity and phenotypic cellular variety. Furthermore, knowledge of such mechanisms is clearly going to be necessary if we are to design a rational stem cell therapy to provide functional neurons to replace those lacking in HSCR.

Neural Crest Origin and Development of ENS Stem Cells

The neural crest (NC) is a specialised transient embryonic tissue, the cells of which emerge from both sides of the neural plate along the cranio-caudal axis of the early embryo. The NC cells then undergo extensive migration and differentiation to form various structures throughout the body including the ENS. The NC cells that become the stem cells for the majority of the ENS originate from the vagal region of the NC and enter the gut in the pharyngeal region, before migrating along the whole length of the gut in a rostrocaudal wave toward the anus [10].

The first ENS stem cells arrive in the foregut on embryonic day 9.5 (E9.5) in the mouse, (approximately E26 in humans) [6, 11, 12]. The leading edge of this rostrocaudal wave in the mouse enters the midgut on E10.5, the caecum on E11.5 and gut colonization is complete by day 14.5 [7, 13]. In human embryos, the stem cells enter the foregut at week 4, and reach the terminal hindgut by week 7 [14]. These cells migrate through the embryonic bowel in chains and the leading cells in the migratory wave provide a scaffold which is then maintained for several hours [15], creating the route that is followed by later cells. Although the migration is very predicable in terms of the population, individual cell analysis shows unpredictable routes [16]. It has also been noted that the net speed of migration of the NC cells in mouse embryos is 35µm per hour [16].

Unlike the CNS, where cell death is the main mechanism of controlling cell number, the driving force in populating the ENS is rapid cell proliferation [17]. A detailed mathematical model, based on vagal NC cell migration, which takes into account the above factors as well as concomitant growth of the bowel during the migration period, has recently been published [18]. Once the NC cells have commenced the colonisation of the cecum, it has been shown that only a small number of stem cells from the leading edge of the wave of migration are sufficient to colonise the entire colon [19, 20]. Although transgenic mouse work has demonstrated that sacral NC cells also colonize the distal hindgut, they do so only after the arrival of the rostrocaudal wave of vagal NCC, and they are unable rescue the HSCR phenotype in the absence of vagal NC cells [21].

FACTORS REGULATING ENS STEM CELL DEVELOPMENT

It has recently become clear that initial specification of the NC stem cell phenotype is controlled by selective expression of a number of transcription factors including snail, Sox -8, -9 and -10, FoxD3 and c-Myc [22]. The events initiating expression of these transcription factors are in turn regulated around the time of delamination of NC cells from the neural plate by a scries of extracellular signalling molecules including bone morphogenetic proteins (BMPs), Wnt and fibroblast growth factors (FGFs) [22]. These signalling molecules often act during the subsequent development of NC cell progeny, but with different developmental outcomes depending on context [23]. Additionally, many other factors and associated signalling pathways have been identified as essential to this process. The majority of factors are detailed below in the text, but in this burgeoning field there are many whose role is less clear or which have not been fully elucidated, and these can be found in Table 1.

Table 1. Other developmental factors involved in the formation of the ENS

Developmental factor	Class	Comments and functions	References
CNTF	Growth factor	Unidentified signalling pathway. Promotes neuronal differentiation.	[57, 59]
Neuregulin/erbB2	Growth factor + receptor	Maintains postnatal ENS neurons via epithelial signalling. Developmental effects unknown.	[102]
ZFHX1B/Sip-1	Transcription factor	May regulate E-cadherin expression. Specific HSCR-mental retardation phenotype in humans.	[85, 103]
Indian hedgehog	Differentiation factor	Involved in later ENS development. Knockout mice have patchy aganglionosis but also other abnormalities of gut development.	[52]
Hoxb3	Homeobox gene	Specific to vagal neural crest cells. Functional effects not characterised.	[104]
Hoxb5	Homeobox gene	ENSC maintenance/self-renewal?	[105]
Hox11L1	Homeobox gene	ENS patterning. Knockout mice develop hyperganglionosis, not hypoganglionosis. Controversy exists over functional effects.	[106-108]
Hand2	Transcription factor	Expression of Hand2 is regulated by BMP2/4 and GDNF. Interacts with Phox2a to produce sympathetic nervous system neurons – may be involved in neuronal specification.	[109, 110]
Foxd3	Transcription factor	Involved in zebrafish ENS development. Significance in human HSCR unknown.	[111]
Trap 100	Receptor complex	Involved in zebrafish ENSC proliferation. Significance in human HSCR unknown.	[112]

GDNF/RET Signalling

Glial cell derived neurotrophic factor (GDNF) is important for the development of subpopulations of both peripheral and central neurons. It is part of a family of signalling molecules that also comprises artemin, neurturin and persephin, all of which are distantly related to transforming growth factor β (TGFβ) [24]. GDNF acts through the RET tyrosine kinase receptor via an accessory molecule, GFRα1; the other members of the GFRα family providing specificity for the binding of artemin, persephin and neurturin [25, 26]. Homozygous Ret knockout mice show complete intestinal aganglionosis (with the exception of the oesophagus as this is derived from somatic neural crest cells), as well as defects of the superior cervical sympathetic ganglia and renal agenesis [27]. An identical phenotype in mice is created in *gdnf-/-* mutants [28-30].

GDNF has been shown in a variety of culture systems to promote NC cell survival, proliferation, differentiation and migration [31-34]. Migrating NC stem cells are initially unresponsive to GDNF [35, 36], and it is only on entering the foregut that they begin to express the RET receptor, and so can be chemotactically attracted by GDNF [36]. Expression of GDNF by mesenchymal tissues is not uniform throughout the developing gut, with both GDNF and endothelin-3 (see below) having maximal concentration in the cecum [37, 38]. As cecal expression of GDNF occurs just prior to arrival of NC cells at the cecum, this observation is consistent with the idea that it plays a role in the directed caudal movement of the cells.

The functions of the other members of the GDNF family (artemin, neurturin and persephin) in the ENS are less well characterised. Murine models which are neurturin- or GFRα2- deficient have a decreased number of neurons in the ENS; the reasons for this remain unclear [39, 40].

Endothelin 3/ENDRB Signalling

Endothelin-3 (EDN3) is one of the three members of the endothelin family of intercellular messenger molecules. Endothelins are ligands for cell surface endothelin receptors A and B.

In the gut EDN3 acts principally via endothelin receptor B (EDNRB), which begins to be expressed by NC cells under control of Sox10 [41] in the neural folds prior to NC stem cell migration [38, 42-44].

In contrast to the total enteric aganglionosis seen in the intestine of mice with GDNF/RET knockouts, EDN3 or EDNRB knockout mice show a lack of ENS that is restricted to the terminal colon and rectum, which is very similar in phenotype to HSCR in humans [45, 46]. Prior to the arrival of NC cells in the embryonic cecum, EDN3 has been shown to promote proliferation of avian NC cells and improve survival of mammalian NC cells [47-49]. However the most striking effect of EDN3, very similar to the HSCR phenotype seen in EDN3 knockouts, is related to its action once NC cells have reached the cecum. At this time point, expression of EDN3 has been located in the mesenchyme of the cecum and proximal colon [44], and EDN3 signalling via the EDNRB receptor has been demonstrated to be necessary for colonization of the distal gut [37, 38]. Given the time-dependent interaction between migrating EDNRB-expressing NC cells and cecal mesenchymal EDN3 expression, it has been proposed that EDN3 is necessary to prevent the premature differentiation of ENS stem cells prior to their colonising the distal part of the gut [34, 35].

Sonic Hedgehog, Neurotrophin and BMP Signaling

Sonic hedgehog (SHH) is initially expressed in the endoderm of the developing vertebrate gut [50, 51], and while the colonisation of the gut by NC stem cells has been shown to be complete in SHH -/- mice, there is abnormal neural plexus formation and placement [52]. SHH has been shown to promote the proliferation and inhibit the differentiation and migration of NC stem cells in response to GDNF [53]. These results suggest a role for SHH and EDN3/EDNRB signalling in "fine tuning" the response of NC stem cells to the RET/GDNF signalling system.

Evidence has been presented that BMP2 and 4 limit the size of the ENS while promoting the development of specific subsets of neurons [54] and ganglion formation [55]. These responses to BMP signalling are, at least in part, due to a structural alteration of neural cell adhesion molecule (NCAM-1) through the addition of polysialic acid [56]. As with sonic hedgehog, animal models deficient in BMP signalling display a mild reduction in the progeny of NC stem cells, indicating a modulating role for BMP signalling.

Neurotrophin-3 (NT3) is the only member of the neurotrophin family that has been shown to have a potential role in early ENS development as it promotes differentiation of ENS precursors in vitro [57], resulting in neurite outgrowth and neuronal differentiation from ganglia in infant rats [58]. Studies in knockout mice lacking either NT3 or its receptor have shown reduced numbers of myenteric and submucosal neurons, whilst overexpression of NT3 leads to increased numbers of myenteric neurons only [59].

Transcription Factors

The transcription factor Phox2b is expressed in all NC derived cells [13, 60]. It is known to regulate the expression of RET as well as enzymes necessary for the production of adrenergic neurotransmitters (tyrosine hydroxylase and dopamine β hydroxylase). Phox2b knockout mice have an identical phenotype to RET deficient mice, with no migration of NC stem cells beyond the foregut [61].

Sox10 and Pax3 act synergistically to promote RET expression, binding adjacent sites within an enhancer [62]. Consequently, defects of the sensory, autonomic and the enteric nervous system have been noted in Sox10 deficient mice, in which failure to colonize the terminal gut results in dilatation similar to that seen with incomplete loss of function mutations of RET [41, 63-65].

In contrast to the effects of the Sox10, Pax3 and Phox2b, mammalian achaete-scute homologue 1 (Mash1) is specifically required for the development of the serotoninergic neuronal lineage from NC stem cells in the gut [66]. Details of the interactions between such transcription factors remain to be established.

Cell Adhesion Molecules and the Extracellular Matrix

β1 integrin expression is essential for gut colonisation and structured ganglion formation by NC stem cells, although in its absence, migration along the gut does occur to a limited extent [67]. While the β1 integrin pathway has not yet been implicated in the pathogenesis of HSCR, much attention has been focused on the extracellular matrix (ECM) molecules that are potential integrin β1 ligands as they have been shown to have profound effects on cell migration in many other systems.

The ECM protein laminin is encountered by migrating NC stem cells within the developing gut and has been shown to affect neurite outgrowth and migration of ENS neurons [68, 69]. Although a 100 kDa laminin binding protein (LBP110) was originally reported to be expressed by stem cells and was suggested to be involved in their development [68], the identity of this protein and any functional role as a laminin receptor has not subsequently been established. While abnormalities in laminin expression have been found in patients with HSCR, it remains unclear whether these are a cause or consequence of the disease [70, 71].

The laminin-related netrins (1 and 3 in mice) and their receptors (Deleted in Colon Cancer [DCC] and neogenin) have been detected in embryonic mouse gut [72]. Significantly, NC stem cells express DCC and were found to migrate radially inwards towards the netrins that are expressed on mucosal epithelium [72]. This study also found that netrins promote the survival and development of NC stem cells. As the initial migrating wave of ENSC colonises the bowel in the outer layer of mesenchyme, the netrin/DCC mechanism helps to explain how the radial patterning of the ENS forms with later formation of the submucosal plexus. These findings are important when considering the generation and clinical transplantation of NC stem cells both because it is necessary to maintain multi-potent, undifferentiated precursor cells in vitro and also to provide an environment capable of producing functional, differentiated cells when transplanted in vivo. A summary of the interactions between the most significant signalling pathways is shown in Figure 1.

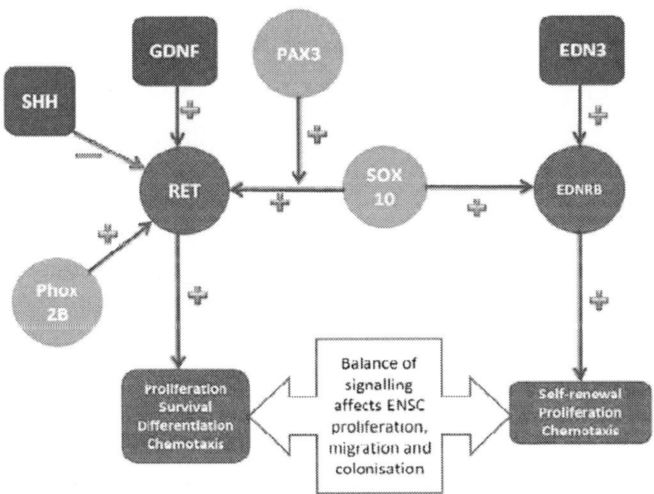

Figure 1. Interactions of the GDNF/RET and EDN3/EDNRB pathways.

The Genetics of ENS Stem Cell Development and HSCR

Perturbations in any of the factors discussed above could potentially affect the development of the ENS, but only a few have known clinical correlations with HSCR. It is likely that the polygenic nature of HSCR is responsible for the lack of simple genetic correlations.

Thus, while mutations in the RET gene are responsible for approximately 50% of familial HSCR cases [73], they can produce a variety of phenotypes in the same family [74, 75]. Linkage studies in affected populations have also identified a non coding mutation in intron 1 of the gene to be associated with HSCR [76], and could explain several features of the complex inheritance pattern of HSCR. This study concluded that RET mutations in coding and/or non-coding sequences are probably a necessary feature of all cases of HSCR. However, the non-coding mutations in isolation are not sufficient for HSCR to occur, and need to be coupled with another mutation of some kind [76].

EDN3 and EDNRB polymorphisms have been described in syndromic and isolated HSCR, although the genetic background is important and penetrance variable [77, 78]. Individuals lacking EDN3 mutations but having decreased levels of EDN3 mRNA expression have also been described [79]. In non-syndromic HSCR, less than 5% of cases appear to be a direct consequence of EDN3/EDNRB mutations [80].

The WS4 variant of Waardenburg-Shah syndrome (HSCR plus partial albinism) has been shown to be related to mutations in SOX10 [63] with patients showing defects in NC cells necessary for both melanocyte and ENS development. WS4 can also be caused by homozygous mutations in EDN3 and ENDRB [63, 81-83]. Another syndrome with a direct genetic link between HSCR and NC stem cell development is Haddad syndrome (central hypoventilation with HSCR), where mutations in Phox2b have been reported as the underlying cause [84, 85].

STRATEGY FOR THE DEVELOPMENT OF A STEM CELL BASED THERAPY FOR HSCR: SOURCES OF STEM CELLS AND THEIR PROPERTIES

The development of a safe and effective stem cell therapy for HSCR requires: a) a source of appropriate stem cells, b) reliable and reproducible methods of isolating these cells, c) evaluation of the properties of these cells, d) effective methods of stem cell transplantation, including the mechanism of delivery and e) control of environmental factors that could affect the outcome of any transplantation.

Sources of Stem Cells

ENS stem cells (ENSC) have been identified in the mouse gut during embryonic development and in the newborn period [20]. ENSC have also been isolated from adult mouse colon [86, 87] and there is evidence that the adult colon may be re-colonized by endogenous ENSC; ablation of a segment of mouse colonic ENS with benzalkonium chloride is followed by restitution via hypertrophy of peripheral myenteric ganglia and, crucially, neurogenesis [88].

Additionally, CNS stem cells (CNSC) remain multipotent, so that cultured neurosphere-derived cells from the subventricular zone of the CNS are capable of migrating along chick neural crest pathways and differentiating into phenotypes similar to NC cell derivatives [89]. A more speculative approach has been investigated by the use of stem cells that are not direct neural progenitors. Thus, attempts have been made to induce the differentiation of embryonic stem cells [90] and bone marrow-derived mesenchymal stem cells towards a neuronal phenotype [72, 91]. The bone marrow cells were induced to differentiate by treatment of with medium conditioned by ENSC [91]. Although the neurons produced expressed neurotransmitter phenotypes found in the ENS including nitric oxide synthase (NOS) and vasoactive intestinal polypeptide (VIP), these transmitters are not restricted to the ENS and therefore any claim for directed differentiation towards a specific enteric neuronal phenotype requires further investigation.

The feasibility of using stem cells that are not the immediate progenitors of the ENS is open to question, as it remains to be seen if the phenotypes of cells derived from them are exactly those of the neural cells required. Furthermore, stem cells – particularly embryonic stem cells – have tumorigenic potential. While the extent of such problems for ENSC is clearly less than for stem cells from other sources, clearly the use of ENSC derived from human embryos is ethically problematic and such cells would also have the disadvantage that their therapeutic use would necessitate an allograft, with consequent immunological problems and the possibility of transmissible infections. These problems would be eliminated by the isolation of ENSC from the normally innervated proximal colon of children affected by HSCR, after diagnosis [92].

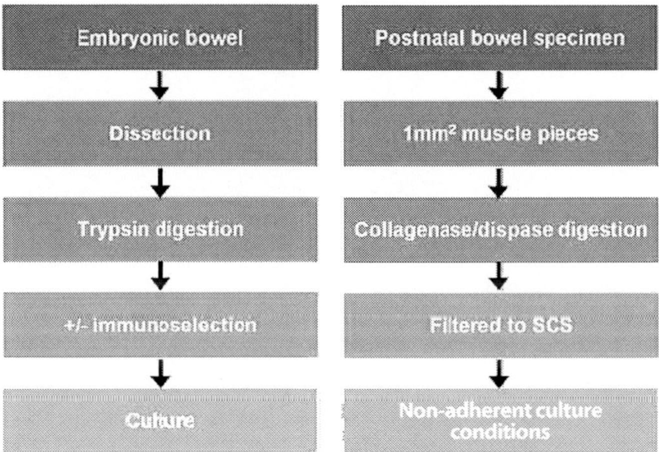

Figure 2. Comparison of methods for isolation of embryonic and postnatal ENSC. Note that the differences in technique arise from the higher connective tissue content of postnatal bowel and the need for longer and more vigorous tissue digestion, and the relatively higher proportion of non-stem cells present in the resulting cell suspension.

Isolation of ENSC

A variety of techniques have been used to isolate ENSC from embryonic and postnatal mice and rats (see Table 2 for a comparison of techniques and Figure 2 which highlights the differences in techniques for generating ENSC from embryonic and postnatal bowel). The varying methods of isolation may result in differing types of cell isolated as it remains unclear if there is topological or chronological heterogeneity in the population of ENSC in the gut. The use of cell isolation based on markers clearly relies on the assumption that all NC stem cells express the marker [93]. However, to obviate this potential problem, use can be made of the property of ENSC to proliferate as neurospheres–a technique developed for the isolation and amplification of CNSC.

Neurospheres are aggregates of stem cells and their progeny, and arise when cells dissociated from neural tissue are cultured in suspension under non-adherent conditions [94]. The advantage of culturing stem cells as neurospheres is that their numbers can be expanded by dissociating the neurosphere and re-culturing the resulting cell suspension: This leads to an exponential increase in the number of neurospheres and hence stem cells and their progeny available for transplantation [95].

Properties of ENSC

Primary neurospheres generated from embryonic mouse bowel contain a heterogeneous group of cells including differentiated neurons and glia, together with a small number of peripherally located cells that are positive for the neural crest-derived stem cell marker p75 (Figure 3; for references, see Table 2).

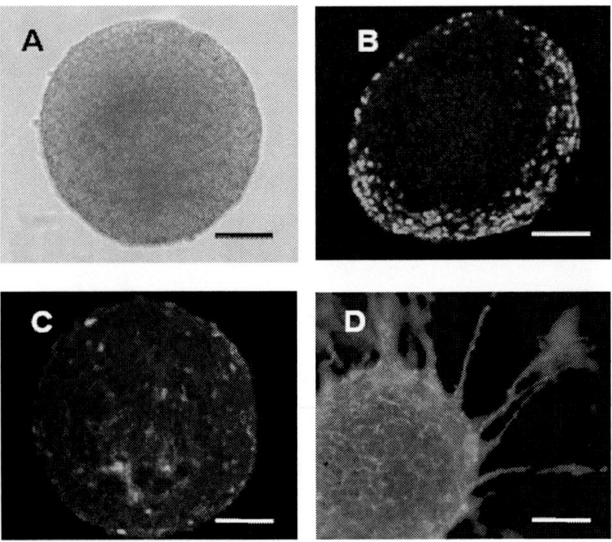

Figure 3. Neurospheres derived from embryonic mouse neural crest cells. A: phase contrast photomicrograph of 28 day old neurosphere. B: immunofluorescence for the neural crest marker p75. C: immunofluorescence for the glial cell marker S100B. D: immunofluorescence for the neuronal cell marker PGP9.5. Scale bars = 200µm.

Following the dissociation of neurospheres into single cell suspensions and subsequent reculture, an exponential rise in neurosphere numbers occurs, with an increase in the proportion of cells positive for p75 [92]. Thus ENSC not only self renew but also appear able to expand their number. BrdU incorporation has shown the most rapidly dividing cells are located peripherally in the neurosphere [92], which is again consistent with the concept of ENSC generating new cells within the neurosphere.

The exact number of ENSC within a primary embryonic neurosphere will obviously vary but experiments using a green fluorescent protein (GFP) retrovirus to transfect dissociated neurospheres indicate that it is about 1-2% of all cells within the neurosphere [20], and this number seems consistent with observed p75 staining within the neurosphere [92].

Table 2. Comparison of methods for the isolation and culture of enteric neural crest stem cell

Species	Age	ENSC Isolation mthod	Culture method	Reference
Rat	Embryonic	Dissociation of whole gut and magnetic micro-bead selection for p75 positive cells	Adherent conditions on collagen and laminin substrate with serum supplemented medium initially	[5, 53, 112]
Mouse	Embryonic	Whole gut dissociation with FACS sorting of RET+ cells	Cultured in explanted E11.5 mouse gut in organ bath	[113]
Rat	Embryonic	Dissociation of whole gut with FACS for p75+/α4 integrin+	Adherent conditions plates coated with PDL and FN with lower PaO2. Colonies, not neurospheres	[96]
Rat	Embryonic and postnatal	As per [96]	As per [96]	[86]
Rat	Embryonic	Dissociation of whole gut with FACS sorting for p75+/α4 integrin positive cells	Adherent conditions plates coated with poly-D-lysine (PDL) and FN. Colonies, not neurospheres	[114]
Mouse	Embryonic	Dissociation of whole gut	Neurospheres grown in culture flask, adherent and mechanically loosened	[115]
Mouse	Embryonic	Whole gut dissociation	Adherent conditions with fibronectin (FN) coated plates	[20]
Mouse	Embryonic and postnatal	Collagenase / dispase dissociation of whole gut, differential sedimentation	Adherent conditions on chamber slides at low density (50-100 cells per 25mm2)	[87]
Mouse	Embryonic	Dissociation of whole gut	Adherent conditions plated coated with PDL and FN, recultured clusters and eventually neurospheres	[53]
Mouse	Embryonic and postnatal	Digestion of cecum only	Adherent conditions.	[92]
Human	Postnatal	Colon	Adherent conditions	[90]

Primary neurospheres begin to form after 5 days in culture and continue to grow for up to 55 days, achieving diameters of approximately 300μm [92]. It has been our experience that the best results for neurosphere propagation are achieved using non-adherent culture conditions and dissociation to form secondary and tertiary neurospheres after 21 days of primary culture (unpublished observations and [92]). Finally, the most important property of embryonic ENSC grown as neurospheres is their potential to produce the full phenotypic range of ENS neurons seen in the adult gut [20, 92, 96].

Age-Related Differences in ENSC Properties

There appear to be significant differences between ENSC isolated in embryonic life and those isolated postnatally. Firstly, the mitotic rate of ENSC derived from postnatal mice is lower than that of embryonic mice [20, 86]. The range of neuronal phenotypes produced by ENSC generated from postnatal tissue would also appear to be restricted. Although NOS, VIP and neuropeptide Y (NPY) neurogenesis appeared normal in postnatal mouse ENSC cultured in adherent conditions, a loss or reduction in ability to form serotoninergic, tyrosine hydroxylase (TH), and dopamine beta-hydroxylase neurons, together with a shift towards gliogenesis was observed [86]. There may also be inter-species differences in the properties of ENSC. When studying ENSC derived from postnatal human bowel we observed that while the cultured human and mouse neurospheres contained cells with similar patterns of neurotransmitter expression, after transplantation into aganglionic embryonic mouse hindgut the postnatal human transplants expressed fewer neuronal phenotypes, as no evidence of immunoreactivity for TH, calcitonin gene related peptide (CGRP) or substance P (SP) was found [92]. However, the presence of CGRP containing neurons within untransplanted neurospheres supports the hypothesis that the progenitor cells are capable of producing cells from the Mash1 independent lineage [97, 98] and therefore have retained their pluripotency. The lack of these neurotransmitters in cells migrating from our transplanted human neurospheres may therefore reflect the absence of appropriate signalling cues in the mouse embryonic bowel, rather than a restriction in neuronal phenotypes. Given the potential advantages of neonatally derived ENSC over embryonic ENSC, the investigation of such potential differences, including the response of ENSC to growth factors is of particular interest and is the subject of ongoing enquiry.

STEM CELL TRANSPLANTATION

Initial attempts at stem cell transplantation into the gut used neurosphere-derived CNSC injected as a single cell suspension into the stomach wall of adult mice which subsequently corrected the deficient neuronal NOS in these animals [99]. Significantly, a functional effect of the transplanted cells was reported, showing increased relaxation in the stomach wall of mice [99].

More extensive work has been performed to assess the characteristics of both mouse and neonatal human ENSC by transplanting them into cultured hindgut explants from day 11.5 mouse embryos. At this timepoint, the wave of migration of NC stem cells has not yet colonised this section of the bowel and the explant therefore normally remains aganglionic [20, 92]. In this model, neurosphere-derived cells have been shown to migrate, colonise the bowel wall, and differentiate into both neurons and glia [92]. The range of neuronal phenotypes produced, as detected by immunofluorescence for various neurotransmitters, is also appropriate for the ENS and includes acetylcholine, VIP, NPY, CGRP [20] and NOS [20, 92]. Examples of a cultured embryonic mouse gut transplanted with human ENS stem cells are shown in Figure 4.

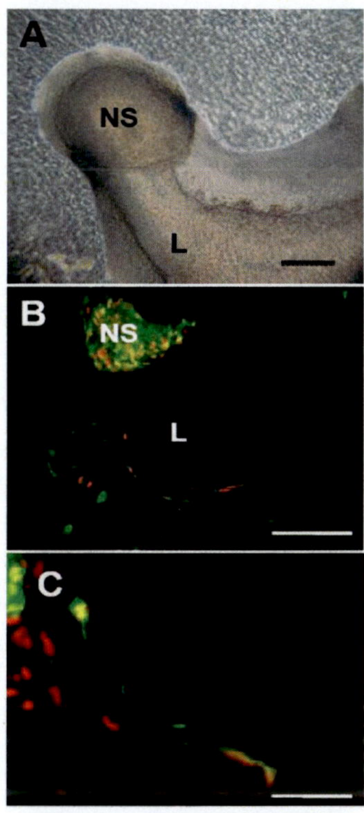

Figure 4. Neurospheres derived from postnatal human bowel transplanted into aganglionic mouse embryonic hindgut. A : phase contrast photomicrograph at day 0. B: immunoflour-escence showing human nuclei (HRNP, red) and neurons (PGP 9.5, green). Note the presence of migrated neurosphere-derived neurons within the bowel wall. C: Detail showing human nuclei (HRNP, red) and glia (S100b, green). L indicates the lumen of the bowel. NS indicates the neurosphere transplant. HRNP=human ribonucleoprotein. (For methods see [90]). Scale bars: A: 200µm; B: 100µm; C: 50µm.

Environmental Factors in the Gut Affecting ENSC Behaviour

The environment that ENSC encounter when transplanted into the gut will play a crucial role in the development and behaviour of their progeny. Furthermore, the response of the gut wall to ENSC and their progeny may well differ in an age-dependent manner. Indeed, the environment into which the cells will be transplanted may be abnormal in HSCR patients. This is potentially the case for extracellular matrix components including laminin [71] or mesenchymal signals including EDN3 [47, 100], both of which have been associated with HSCR. The use of supplemental growth factors is one possible way in which these signals could be delivered, avoiding the inherent problems of the abnormal tissue.

A further problem is that transplanted ENSC may be defective if derived from an autologous source (e.g. lacking the RET or ENDRB receptors). However, based on genetic studies on HSCR, any deficits in gene/receptor function will be small and multifactorial, and may not be evident over the short distances transplanted cells must migrate. Furthermore, while it might be expected that ENSC from patients affected with RET mutations would be unable to colonise aganglionic bowel, it has been shown that ENSC derived from mice with the miRet[51] mutation are capable of forming neurospheres and differentiating into both neurons and glia in culture [20]. In theory this could allow auto-transplantation of an HSCR patient's own ENSC into aganglionic bowel.

It should be noted that one of the reasons for the relatively poor outcome after surgery for HSCR is likely to be the amount of aganglionic tissue left in the patient, including the anal sphincter and a few centimeters of rectum [101]. Re-innervation of this short segment of bowel with implanted autologous ENSC generated from tissue removed during surgery is the most feasible potential stem cell therapy for HSCR.

CONCLUSION

There is now a considerable body of knowledge on the developmental behaviour of ENS stem cells and the mechanisms behind HSCR. This knowledge is now being utilized in the search for a successful stem cell-based therapy for HSCR. What form this will take is currently not clear, as there are a wide variety of stem cell sources that could be utilized for the recolonisation of aganglionic bowel by the ENS, numerous methods of isolation, several different phenotypes of neuron that can be generated, and a variety of transplantation techniques already in the literature. The next stage in ENS research will hopefully be to use these techniques to demonstrate a functional change in HSCR-like bowel from a transplantation of stem cells.

ACKNOWLEDGMENTS

We thank the Royal College of Surgeons (England), CORE/the Digestive Disorders Foundation UK, and Action Medical Research UK for financial support.

REFERENCES

[1]   Bodian M, Carter CO. 1963 A family study of Hirschsprungs disease. *Ann. Hum. Genet.* 26(3): 261-77.

[2]   Swenson O, Bill AH. 1948 Resection of rectum and rectosigmoid with preservation of the sphincter for benign spastic lesions producing megacolon - an experimental study. *Surgery* 24(2): 212-20.

[3]   Baillie CT, Kenny SE, Rintala RJ, Booth JM, Lloyd DA. 1999 Long-term outcome and colonic motility after the Duhamel procedure for Hirschsprung's disease. *J. Pediatr. Surg.* 34(2): 325-29.

[4]   Tsuji H, Spitz L, Kiely EM, Drake DP, Pierro A. 1999 Management and long-term follow-up of infants with total colonic aganglionosis. *J. Pediatr. Surg.* 34(1): 158-61.

[5]   Chalazonitis A, Rothman TP, Chen J, Gershon MD. 1998 Age-dependent differences in the effects of GDNF and NT-3 on the development of neurons and glia from neural crest-derived precursors immunoselected from the fetal rat gut: expression of GFRalpha-1 in vitro and in vivo. *Dev. Biol.* 204(2): 385-406.

[6]   Newgreen D, Young HM. 2002 Enteric nervous system: Development and developmental disturbances - Part 2. *Pediatr. Dev. Pathol.* 5(4): 329-49.

[7]   Newgreen D, Young HM. 2002 Enteric nervous system: Development and developmental disturbances - Part 1. *Pediatr. Dev. Pathol.* 5(3): 224-47.

[8]   Grundy D, Schemann M. 2005 Enteric nervous system. *Curr. Opin. Gastroen* .21(2): 176-82.

[9]   Ruhl A, Nasser Y, Sharkey KA. 2004 Enteric glia. *Neurogastroenterol Motil* 16: 44-49.

[10]   Yntema C, Hammond W. 1954 The origin of intrinsic ganglia of trunk viscera from vagal neural crest in the chick embryo. *J. Comp. Neurol.* 101: 515-41.

[11]   Le Douarin N, Teillet M. 1973 The migration of neural crest cells to the wall of the digestive tract in avian embryo. *J. Embryol. Exp. Morphol.* 30: 31-48.

[12]   Kapur RP, Yost C, Palmiter RD. 1992 A transgenic model for studying development of the enteric nervous-system in normal and aganglionic mice. *Development* 116(1): 167-and.

[13]   Young HM, Hearn CJ, Ciampoli D, Southwell BR, Brunet JF, Newgreen DF. 1998 A single rostrocaudal colonization of the rodent intestine by enteric neuron precursors is revealed by the expression of Phox2b, Ret, and p75 and by explants grown under the kidney capsule or in organ culture. *Dev. Biol.* 202(1): 67-84.

[14]   Wallace AS, Burns AJ. 2005 Development of the enteric nervous system, smooth muscle and interstitial cells of Cajal in the human gastrointestinal tract. *Cell Tissue Res.* 319(3): 367-82.

[15]   Young HM, Anderson RB, Anderson CR. 2004 Guidance cues involved in the development of the peripheral autonomic nervous system. *Autonomic Neuroscience Basic and Clinical* 112(1-2): 1-14.

[16]   Young HM, Bergner AJ, Anderson RB, Enomoto H, Milbrandt J, Newgreen DF, Whitington PM. 2004 Dynamics of neural crest-derived cell migration in the embryonic mouse gut. *Dev. Biol.* 270(2): 455-73.

[17]   Gianino S, Grider JR, Cresswell, J Enomoto, H Heuckeroth RO. 2003 GDNF availability determines enteric neuron number by controlling precursor proliferation. *Development* 130(10): 2187-98.

[18]   Simpson MJ, Landman KA, Newgreen DF. 2006 Chemotactic and diffusive migration on a nonuniformly growing domain: numerical algorithm development and applications. *J Comput. Appl. Math.* 192(2): 282-300.

[19]   Sidebotham EL, Woodward MN, Kenny SE, Lloyd DA, Vaillant CR, Edgar DH. 2002 Localization and endothelin-3 dependence of stem cells of the enteric nervous system in the embryonic colon. *J. Pediatr. Surg.* 37(2): 145-50.

[20]   Bondurand N, Natarajan D, Thapar N, Atkins C, Pachnis V. 2003 Neuron and glia generating progenitors of the mammalian enteric nervous system isolated from foetal and postnatal gut cultures. *Development* 130(25): 6387-400.

[21]   Kapur RP. 2000 Colonization of the murine hindgut by sacral crest-derived neural precursors: Experimental support for an evolutionarily conserved model. *Dev. Biol.* 227(1): 146-55.

[22]   Barembaum M, Bronner-Fraser M. 2005 Early steps in neural crest specification. *Semin. Cell Dev. Biol* .16(6): 642-46.

[23]   Raible DW, Ragland JW. 2005 Reiterated Wnt and BMP signals in neural crest development. *Semin. Cell Dev. Biol.* 16(6): 673-82.

[24] Massague J. 1996 TGF beta signaling: Receptors, transducers, and mad proteins. *Cell* 85(7): 947-50.

[25] Jing SQ, Wen DZ, Yu YB, Holst PL, Luo Y, Fang M, Tamir R, Antonio L, Hu Z, Cupples R, Louis JC, Hu S, Altrock BW, Fox GM. 1996 GDNF-induced activation of the Ret protein tyrosine kinase is mediated by GDNFR-alpha, a novel receptor for GDNF. *Cell* 85(7): 1113-24.

[26] Treanor JJS, Goodman L, deSauvage F, Stone DM, Poulsen KT, Beck CD, Gray C, Armanini MP, Pollock RA, Hefti F, Phillips HS, Goddard A, Moore MW, BujBello A, Davies AM, Asai N Takahashi M, Vandlen R, Henderson CE, Rosenthal A. 1996 Characterization of a multicomponent receptor for GDNF. *Nature* 382(6586): 80-83.

[27] Schuchardt A, Dagati V, Larssonblomberg L, Costantini F, Pachnis V. 1994 Defects in the kidney and enteric nervous-system of mice lacking the tyrosine kinase receptor Ret. *Nature* 367(6461): 380-83.

[28] Moore MW, Klein RD, Farinas I, Sauer H, Armanini M, Phillips H, Reichardt LF, Ryan AM, CarverMoore K, Rosenthal A. 1996 Renal and neuronal abnormalities in mice lacking GDNF. *Nature* 382(6586): 76-79.

[29] Pichel JG, Shen LY, Sheng HZ, Granholm AC, Drago J, Grinberg A, Lee EJ, Huang SP, Saarma M, Hoffer BJ, Sariola H, Westphal H. 1996 Defects in enteric innervation and kidney development in mice lacking GDNF. *Nature* 382(6586): 73-76.

[30] Sanchez MP, SilosSantiago I, Frisen J, He B Lira SA, Barbacid M. 1996 Renal agenesis and the absence of enteric neurons in mice lacking GDNF. *Nature* 382(6586): 70-73.

[31] Ebendal T, Tomac A, Hoffer BJ, Olson L. 1995 Glial-Cell Line-Derived Neurotrophic Factor stimulates fiber formation and survival in cultured neurons from peripheral autonomic ganglia. *J. Neurosci. Res.* 40(2): 276-84.

[32] Matheson CR, Carnahan J, Urich JL, Bocangel, D Zhang TJ, Yan Q. 1997 Glial cell line-derived neurotrophic factor (GDNF) is a neurotrophic factor for sensory neurons: Comparison with the effects of the neurotrophins. *J. Neurobiol.* 32(1): 22-32.

[33] Molliver DC, Wright DE, Leitner ML, Parsadanian AS, Doster K, Wen D, Yan Q, Snider WD. 1997 IB4-binding DRG neurons switch from NGF to GDNF dependence in early postnatal life. *Neuron* 19(4): 849-61.

[34] Wu JJ, Chen JX, Rothman TP, Gershon MD. 1999 Inhibition of in vitro enteric neuronal development by endothelin-3: mediation by endothelin B receptors. *Development* 126(6): 1161-73.

[35] Hearn CJ, Murphy M, Newgreen D. 1998 GDNF and ET-3 differentially modulate the numbers of avian enteric neural crest cells and enteric neurons in vitro. *Dev. Biol.* 197(1): 93-105.

[36] Young HM, Hearn CJ, Farlie PG, Canty AJ, Thomas PQ, Newgreen DF. 2001 GDNF is a chemoattractant for enteric neural cells. *Dev. Biol.* 229(2): 503-16.

[37] Shin MK, Levorse JM, Ingram RS, Tilghman SM. 1999 The temporal requirement for endothelin receptor-B signalling during neural crest development. *Nature* 402(6761): 496-501.

[38] Woodward MN, Kenny SE, Vaillant C, Lloyd DA, Edgar DH. 2000 Time-dependent effects of endothelin-3 on enteric nervous system development in an organ culture model of Hirschsprung's disease. *J. Pediatr. Surg.* 35(1): 25-9.

[39] Heuckeroth RO, Enomoto H, Grider JR, Golden JP, Hanke JA, Jackman A, Molliver DC, Bardgett ME, Snider WD, Johnson EM, Milbrandt J. 1999 Gene targeting reveals a

critical role for neurturin in the development and maintenance of enteric, sensory, and parasympathetic neurons. *Neuron* 22(2): 253-63.

[40]   Rossi J, Luukko K, Poteryaev D, Laurikainen A, Sun YF, Laakso T, Eerikainen S, Tuominen R, Lakso M, Rauvala H, Arumae U, Pasternack M, Saarma M, Airaksinen MS. 1999 Retarded growth and deficits in the enteric and parasympathetic nervous system in mice lacking GFR alpha 2, a functional neurturin receptor. *Neuron* 22(2): 243-52.

[41]   Britsch S, Goerich DE, Riethmacher D, Peirano RI, Rossner M, Nave KA, Birchmeier C, Wegner M. 2001 The transcription factor Sox10 is a key regulator of peripheral glial development. *Genes Dev.* 15(1): 66-78.

[42]   Nataf V, Lecoin L, Eichmann A, LeDouarin NM. 1996 Endothelin-B receptor is expressed by neural crest cells in the avian embryo. *Proc. Natl. Acad. Sci. USA* 93(18): 9645-50.

[43]   Brand M, Le Moullec JM, Corvol P, Gasc JM. 1998 Ontogeny of endothelins-1 and -3, their receptors, and endothelin converting enzyme-1 in the early human embryo. *J. Clin. Invest.* 101(3): 549-59.

[44]   Leibl MA, Ota T, Woodward MN, Kenny SE, Lloyd DA, Vaillant CR, Edgar DH. 1999 Expression of endothelin 3 by mesenchymal cells of embryonic mouse caecum. *Gut* 44(2): 246-52.

[45]   Baynash AG, Hosoda K, Giaid A, Richardson JA, Emoto N, Hammer RE, Yanagisawa M. 1994 Interaction of endothelin-3 with endothelin-B receptor is essential for development of epidermal melanocytes and enteric neurons. *Cell* 79(7): 1277-85.

[46]   Hosoda K, Hammer RE, Richardson JA, Baynash AG, Cheung JC, Giaid A, Yanagisawa M. 1994 Targeted and Natural (Piebald-Lethal) Mutations of Endothelin-B receptor gene produce megacolon aAssociated with spotted Coat color in mice. *Cell* 79(7): 1267-76.

[47]   Nagy N, Goldstein AM. 2006 Endothelin-3 regulates neural crest cell proliferation and differentiation in the hindgut enteric nervous system. *Dev. Biol.* 293(1): 203-17.

[48]   Barlow A, de Graaff E, Pachnis V. 2003 Enteric nervous system progenitors are coordinately controlled by the G protein-coupled receptor EDNRB and the receptor tyrosine kinase RET. *Neuron.* 40(5): 905-16.

[49]   Kruger GM, Mosher JT, Tsai TH, Yeager KJ, Iwashita T, Gariepy CE, Morrison SJ. 2003 Temporally distinct requirements for endothelin receptor B in the generation and migration of gut neural crest stem cells. *Neuron* 40(5): 917-29.

[50]   Bitgood MJ, McMahon AP. 1995 Hedgehog and Bmp genes are coexpressed at many diverse sites of cell-cell interaction in the mouse embryo. *Dev. Biol.* 172(1): 126-38.

[51]   Roberts D, Johnson RL, Burke AC, Nelson CE, Morgan BA, Tabin C. 1995 Sonic hedgehog is an endodermal dignal inducing Bmp-4 and hox genes during induction and regionalization of the chick hindgut. *Development* 121(10): 3163-74.

[52]   Ramalho-Santos M, Melton DA, McMahon AP. 2000 Hedgehog signals regulate multiple aspects of gastrointestinal development. *Development* 127(12): 2763-72.

[53]   Fu M, Lui VC, Sham MH, Pachnis V, Tam PK. 2004 Sonic hedgehog regulates the proliferation, differentiation, and migration of enteric neural crest cells in gut. *J. Cell Biol.* 166(5): 673-84.

[54]   Chalazonitis A, D'Autreaux F, Guha U, Pham TD, Faure C, Chen JJ, Roman D, Kan LX, Rothman TP, Kessler JA, Gershon MD. 2004 Bone morphogenetic protein-2 and-4

limit the number of enteric neurons but promote development of a TrkC-expressing neurotrophin-3-dependent subset. *J. Neurosci.* 24(17): 4266-82.

[55]   Goldstein AM, Brewer KC, Doyle AM, Nagy N, Roberts DJ. 2005 BMP signaling is necessary for neural crest cell migration and ganglion formation in the enteric nervous system. *Mech. Dev.* 122(6): 821-33.

[56]   Fu M, Vohra BPS, Wind D, Heuckeroth RO. 2006 BMP signaling regulates murine enteric nervous system precursor migration, neurite fasciculation, and patterning via altered Ncam1 polysialic acid addition. *Dev. Biol.* 299(1): 137-50.

[57]   Chalazonitis A, Rothman TP, Chen JX, Vinson EN, MacLennan AJ, Gershon MD. 1998 Promotion of the development of enteric neurons and glia by neuropoietic cytokines: Interactions with neurotrophin-3. *Dev. Biol.* 198(2): 343-65.

[58]   Saffrey MJ, Wardhaugh T, Walker T, Daisley J, Silva AT. 2000 Trophic actions of neurotrophin-3 on postnatal rat myenteric neurons in vitro. *Neurosci. Lett* .278(3): 133-36.

[59]   Chalazonitis A, Pham TD, Rothman TP, DiStefano PS, Bothwell M, Blair-Flynn J, Tessarollo L, Gershon MD. 2001 Neurotrophin-3 is required for the survival-differentiation of subsets of developing enteric neurons. *J. Neurosci.* 21(15): 5620-36.

[60]   Canty AJ. 1999 Expression of Ret-, p75(NTR)-, Phox2a-, Phox2b-, and tyrosine hydroxylase-immunoreactivity by undifferentiated neural crest-derived cells and different classes of enteric neurons in the embryonic mouse gut. *Dev. Dyn.* 216(2): 137-52.

[61]   Pattyn A, Morin X, Cremer H, Goridis C, Brunet JF. 1999 The homeobox gene Phox2b is essential for the development of autonomic neural crest derivatives. *Nature* 399(6734): 366-70.

[62]   Lang D, Chen F, Milewski R, Li, J Lu MM, Epstein JA. 2000 Pax3 is required for enteric ganglia formation and functions with Sox10 to modulate expression of c-ret. *J. Clin. Invest.* 106(8): 963-71.

[63]   Southard-Smith EM, Kos L, Pavan WJ. 1998 Sox10 mutation disrupts neural crest development in DOM Hirschsprung mouse model. *Nat. Genet.* 18(1): 60-64.

[64]   Herbarth B, Pingault V, Bondurand N, Kuhlbrodt K, Hermans-Borgmeyer I, Puliti A, Lemort N, Goossens M, Wegner M. 1998 Mutation of the Sry-related Sox10 gene in Dominant megacolon, a mouse model for human Hirschsprung disease. *Proc. Natl. Acad. Sci. USA* 95(9): 5161-65.

[65]   Kapur RP. 1999 Early death of neural crest cells is responsible for total enteric aganglionosis in Sox10(Dom)/Sox10(Dom) mouse embryos. *Pediatr. Dev. Pathol.* 2(6): 559-69.

[66]   Sommer L, Shah N, Rao M, Anderson DJ. 1995 The cellular function of MASH1 in autonomic neurogenesis. *Neuron* 15(6): 1245-58.

[67]   Breau MA, Pietri T, Eder O, Blanche M, Brakebusch C, Fassler R, Thiery JP, Dufour S. 2006 Lack of beta 1 integrins in enteric neural crest cells leads to a Hirschsprung-like phenotype. *Development* 133(9): 1725-34.

[68]   Pomeranz HD, Sherman DL, Smalheiser NR, Tennyson VM, Gershon MD,.1991 Expression of a neurally related laminin binding-protein by neural crest-derived cells that colonize the gut. *J. Comp. Neurol.* 313(4): 625-42.

[69]   Miner JH, Yurchenco. PD 2004 Laminin functions in tissue morphogenesis. *Annu. Rev. Cell Dev. Biol.* 20: 255-84.

[70] Parikh DH, Tam PKH, Vanvelzen D, Edgar D. 1992 Abnormalities in the distribution of laminin and collagen type-IV in Hirschsprungs disease. *Gastroenterology* 102(4): 1236-41.

[71] Alpy F, Ritie L, Jaubert F, Becmeur F, Mechine-Neuville A, Lefebvre O, Arnold C, Sorokin L, Kedinger M, Simon-Assmann P. 2005 The expression pattern of laminin isoforms in Hirschsprung disease reveals a distal peripheral nerve differentiation. *Hum. Pathol.* 36(10): 1055.

[72] Jiang Y, Liu MT, Gershon MD. 2003 Netrins and DCC in the guidance of migrating cells in the developing bowel and neural crest-derived pancreas. *Dev. Biol.* 258(2): 364-84.

[73] Chakravati A. Hirschsprung's *Disease The metabolic and moleular basis of inherited diseases,*.ed. C Scrivener, A Beaudet, D Valle, W Sly, B Childs, K Kinzler, and B Vogelstein 2001 New York: McGraw-Hill.

[74] Edery P, Pelet A, Mulligan LM, Abel L, Attie T, Dow E, Bonneau D, David A, Flintoff W, Jan D, Journel H, Lacombe D, Lemerrer M, Meijers C, Parent P, Philip N, Plauchu H, Sarda P, Verloes A, Nihoulfekete C, Williamson R, Ponder BAJ, Munnich A, Lyonnet S. 1994 Long-segment and short segment familial Hirschsprungs-disease - Variable clinical expression at the Ret locus. *J. Med. Genet.* 31(8): 602-06.

[75] Romeo G, Ronchetto P, Luo Y, Barone V, Seri M, Ceccherini I, Pasini B, Bocciardi R, Lerone M, Kaariainen H, Martucciello G. 1994 Point mutations affecting the tyrosine kinase domain of the Ret protooncogene in Hirschprungs disease. *Nature* 367(6461): 377-78.

[76] Emison ES, McCallion AS, Kashuk CS, Bush RT, Grice E, Lin S, Portnoy ME, Cutler DJ, Green ED, Chakravarti A. 2005 A common sex-dependent mutation in a RET enhancer underlies Hirschsprung disease risk. *Nature* 434(7035): 857-63.

[77] Puffenberger EG, Hosoda K, Washington SS, Nakao K, Dewit D, Yanagisawa M, Chakravarti A. 1994 A missense mutation of the endothelin-B receptor gene in multigenic Hirschsprungs-disease. *Cell* 79(7): 1257-66.

[78] Kusafuka T, Puri P. 1998 Genetic aspects of Hirschsprung's Disease. *Semin. Pediatr. Surg.* 7: 148-55.

[79] Kenny SE, Hofstra RM, Buys CH, Vaillant CR, Lloyd DA, Edgar DH. 2000 Reduced endothelin-3 expression in sporadic Hirschsprung disease. *Br. J. Surg.* 87(5): 580-5.

[80] Brooks AS, Oostra BA, Hofstra RMW. 2005 Studying the genetics of Hirschsprung's disease: unraveling an oligogenic disorder. *Clin. Genet.* 67(1): 6-14.

[81] Hofstra RMW, Valdenaire O, Arch E, Osinga J, Kroes H, Loffler BM, Hamosh A, Meijers C, Buys C. 1999 A loss-of-function mutation in the endothelin-converting enzyme 1 (ECE-1) associated with Hirschsprung disease, cardiac defects, and autonomic dysfunction. *Am. J. Hum. Genet* 64(1): 304-08.

[82] Pingault V, Bondurand N, Kuhlbrodt K, Goerich DE, Prehu MO, Puliti A, Herbarth B, Hermans-Borgmeyer I, Legius E, Matthijs G, Amiel J, Lyonnet S, Ceccherini I, Romeo G, Smith JC, Read AP, Wegner M, Goossens M. 1998 SOX10 mutations in patients with Waardenburg-Hirschsprung disease. *Nat. Genet.* 18(2): 171-73.

[83] Southard-Smith EM, Angrist M, Ellison JS, Agarwala R, Baxevanis AD, Chakravarti A, Pavan WJ. 1999 The Sox10(Dom) mouse: Modeling the genetic variation of Waardenburg-Shah (WS4) syndrome. *Genome Res.* 9(3): 215-25.

[84] Shibata C, Sasaki I, Matsuno S, Mizumoto A, Itoh Z. 1991 Colonic motility in innervated and extrinsically denervated loops in dogs. *Gastroenterology* 101(6): 1571-78.

[85] Amiel J, Lyonnet S. 2001 Hirschsprung disease, associated syndromes, and genetics: a review. *J. Med. Genet.* 38(11): 729-39.

[86] Kruger GM, Mosher JT, Bixby S, Joseph N, Iwashita T, Morrison SJ. 2002 Neural crest stem cells persist in the adult gut but undergo changes in self-renewal, neuronal subtype potential, and factor responsiveness. *Neuron* 35(4): 657-69.

[87] Suarez-Rodriguez R, Belkind-Gerson J. 2004 Cultured nestin-positive cells from postnatal mouse small bowel differentiate ex vivo into neurons, glia, and smooth muscle. *Stem Cells* 22(7): 1373-85.

[88] Hanani M, Ledder O, Yutkin V, Abu-Dalu R, Huang TY, Hartig W, Vannucchi MG, Faussone-Pellegrin MS. 2003 Regeneration of myenteric plexus in the mouse colon after experimental denervation with benzalkonium chloride. *J. Comp. Neurol.* 462(3): 315-27.

[89] Durbec P, Rougon G. 2001 Transplantation of mammalian olfactory progenitors into chick hosts reveals migration and differentiation potentials dependent on cell commitment. *Mol. Cell Neurosci.* 17: 561-76.

[90] van Leeuwen K, Teitelbaum DH, Elhalaby EA, Coran AG. 2000 Long-term follow-up of redo pull-through procedures for Hirschsprung's disease: Efficacy of the endorectal pull-through. *J. Pediatr. Surg.* 35(6): 829-33.

[91] Bossolasco P, Cova L, Calzarossa C, Rimoldi SG, Borsotti C, Deliliers GL, Silani V, Soligo D, Polli E. 2005 Neuro-glial differentiation of human bone marrow stem cells in vitro. *Exp. Neurol.* 193(2): 312-25.

[92] Almond S, Lindley RM, Kenny S,E Connell MG, Edgar DH. 2006 Characterisation and transplantation of enteric nervous system progenitor cells. *Gut*: gut.2006.doi: 094565.

[93] Young HM, Turner KN, Bergner AJ. 2005 The location and phenotype of proliferating neural-crest-derived cells in the developing mouse gut. *Cell Tissue Res.* 320(1): 1-9.

[94] Reynolds BA, Weiss S. 1996 Clonal and population analyses demonstrate that an EGF-responsive mammalian embryonic CNS precursor is a stem cell. *Dev. Biol.* 175(1): 1.

[95] Campos LS. 2004 Neurospheres: Insights biology into neural stem cell biology. *J. Neurosci. Res.* 78(6): 761-69.

[96] Bixby S, Kruger GM, Mosher JT, Joseph NM, Morrison SJ/ 2002 Cell-intrinsic differences between stem cells from different regions of the peripheral nervous system regulate the generation of neural diversity. *Neuron* 35(4): 643-56.

[97] Li ZS, Pham TD, Tamir H, Chen JJ, Gershon MD/ 2004 Enteric dopaminergic neurons: Definition, developmental lineage, and effects of extrinsic denervation. *J. Neurosci.* 24(6): 1330-39.

[98] Pham TD, Gershon MD, Rothman TP. 1991 Time of origin of neurons in the murine enteric nervous-system - Sequence in relation to phenotype. *J. Comp. Neurol.* 314(4): 789-98.

[99] Micci MA, Kahrig KM, Pasricha PJ. 2005 Nitrinergic re-innervation and improvement of gastric function in nNOS-/- mice by transplantation of CNS-derived stem cells. *Gastroenterology* 128(4): A106-A06.

[100] Bondurand N, Natarajan D, Barlow A, Thapar N, Pachnis V. 2006 Maintenance of mammalian enteric nervous system progenitors by SOX10 and endothelin 3 signalling. *Development* 133(10): 2075-86.

[101] Swenson O. 2002 Hirschsprung's disease: A review. *Pediatrics* 109(5): 914-18.

[102] Crone SA, Negro A, Trumpp A, Giovannini M, Lee KF. 2003 Colonic epithelial expression of ErbB2 is required for postnatal maintenance of the enteric nervous system. *Neuron* 37(1): 29-40.

[103] Van de Putte T, Maruhashi M, Francis A, Nelles L, Kondoh H, Huylebroeck D, Higashi Y. 2003 Mice lacking Zfhx1b, the gene that codes for Smad- interacting protein-1, reveal a role for multiple neural crest cell defects in the etiology of Hirschsprung disease-mental retardation syndrome. *Am. J. Hum. Genet.* 72(2): 465-70.

[104] Tassabehji M, Read AP, Newton VE, Harris R, Balling R, Gruss P, Strachan T. 1992 Waardenburg Syndrome patients have mutations in the human homolog of the Pax-3 paired box gene. *Nature* 355(6361): 635-36.

[105] Fu M, Lui VCH, Sham MH, Cheung ANY, Tam PKH. 2003 HOXB5 expression is spatially and temporarily regulated in human embryonic gut during neural crest cell colonization and differentiation of enteric neuroblasts. *Dev. Dyn.* 228(1): 1-10.

[106] Shirasawa S, Yunker AMR, Roth KA, Brown GA, Horning S, Korsmeyer SJ. 1997 Enx (Hox11L1)-deficient mice develop myenteric neuronal hyperplasia and megacolon. *Nat. Med.* 3(6): 646-50.

[107] Parisi MA, Baldessari AE, Iida MHK, Clarke C,M, Doggett B, Shirasawa S, Kapur RP. 2003 Genetic background modifies intestinal pseudo-obstruction and the expression of a reporter gene in Hox11L1-/- mice. *Gastroenterology* 125(5): 1428-40.

[108] Kapur RP, Clarke CM, Doggett B, Taylor BE, Baldessari A, Parisi MA, Howe DG. 2005 Hox11L1 Expression by precursors of enteric smooth muscle: An alternative explanation for megacecum in Hox11L1 Mice. *Pediatr. Dev. Pathol.* 8(2): 148.

[109] Gao Y, Li GC, Zhang XS, Xu Q, Guo ZT, Zheng BJ, Li P, Li GW. 2001 Primary transanal rectosigmoidectomy for Hirschsprung's disease: Preliminary results in the initial 33 cases. *J. Pediatr. Surg.* 36(12): 1816-19.

[110] Wu XD, Howard MJ. 2002 Transcripts encoding HAND genes are differentially expressed and regulated by BMP4 and GDNF in developing avian gut. *Gene Expr.* 10(5-6): 279-93.

[111] Lister JA, Cooper C, Nguyen K, Modrell M, Grant K, Raible DW. 2006 Zebrafish Foxd3 is required for development of a subset of neural crest derivatives. *Dev. Biol.* 290(1): 92-104.

[112] Shepherd IT, Pietsch J, Elworthy S, Kelsh RN, Raible DW. 2004 Roles for GFR alpha 1 receptors in zebrafish enteric nervous system development. *Development* 131(1): 241-49.

[113] Natarajan D, Grigoriou M, MarcosGutierrez CV, Atkins C, Pachnis V. 1999 Multipotential progenitors of the mammalian enteric nervous system capable of colonising aganglionic bowel in organ culture. *Development* 126(1): 157-68.

[114] Iwashita T, Kruger GM, Pardal R, Kiel MJ, Morrison SJ. 2003 Hirschsprung disease is linked to defects in neural crest stem cell function. *Science* 301(5635): 972-76.

[115] Schafer KH, Hagl CI, Rauch U. 2003 Differentiation of neurospheres from the enteric nervous system. *Pediatr. Surg. Int.* 19(5): 340-44.

In: Stem Cell Research Progress
Editor: Prasad S. Koka, pp. 159-179

ISBN: 978-1-60456-308-5
© 2008 Nova Science Publishers, Inc.

Chapter 10

THE ROLE OF STEM CELLS IN REPAIR OF ACUTE LUNG INJURY

Ellen L. Burnham[*]

University of Colorado at Denver Health Sciences Center, Denver, Colorado

ABSTRACT

The acute respiratory distress syndrome (ARDS) and acute lung injury (ALI) are common diagnoses in intensive care patients who require mechanical ventilation. They are acute in onset and typically fulminant in nature. No medical therapy has been found to improve survival from these disorders despite years of investigation and several clinical trials. Mortality for ARDS continues to approximate 40%. The alveolar epithelium, composed of types I and II cells, is a prime target for damage in ALI/ARDS, as is the pulmonary vasculature. Both the pulmonary epithelium and endothelium are injured diffusely. The development of pulmonary hypertension (PHTN) may be observed as well. Given the lack of available therapies for ALI/ARDS, investigators and clinicians alike are eager to explore new treatment modalities directed at repair of lung damaged in these disorders. As a result, the role of stem cells is receiving increasing attention. This review will focus on the role of stem cells in the repair of lung injury, specifically those studies that focus on the regeneration of lung epithelium and endothelium. Potential therapies for pulmonary hypertension will be highlighted as well. Investigations that have been performed in animal models will be described, and how these may be applicable to human disease.

---

[*] Address Correspondence to: Ellen L. Burnham, MD. University of Colorado Health Sciences Center, Division of Pulmonary Sciences and Critical Care Medicine, 4200 E. Ninth Ave. Bldg 1, C272, Room 5525. Denver, CO 80262. Phone 303-315-1111 Fax 303-315-1356. Email: Ellen.Burnham@uchsc.edu

WHAT IS THE ACUTE RESPIRATORY DISTRESS SYNDROME AND ACUTE LUNG INJURY?

The acute respiratory distress syndrome (ARDS), or its less severe form, acute lung injury (ALI) represent common diagnoses in intensive care patients who require mechanical ventilation [1]. A variety of risk factors can predispose patients to developing ALI/ARDS, including disease processes that affect the lung directly, such as pneumonia, or those that affect the lung indirectly, such as sepsis. Despite years of investigation and several clinical trials [2-4], no medical therapy has been demonstrated to lead to improved mortality from these disorders, and in fact mortality for ARDS continues to range from 30 to 50%. At the present time, a mechanical ventilation strategy is the only therapy known to benefit patients in terms of outcomes, including mortality and number of ventilator-free days [5].

Alveolar epithelium is a prime target for damage in ALI/ARDS. In the normal, undamaged lung, alveolar epithelial type I (AEI) and alveolar epithelial type II (AEII) cells line the alveolus, the primary area for gas exchange. AEI cells are presumed to be terminally differentiated, and as such are particularly susceptible to damage as they cannot divide and repair themselves after injury. AEII cells, normally responsible for synthesizing and secreting surfactant, can divide and differentiate, and have been presumed to be the precursor cells for AEI cells in situations of injury [6]. In ALI/ARDS, the alveolar epithelium becomes diffusely injured [1]. During the acute (days 1 though 3) phase of this disorder, airspaces of patients with ALI/ARDS are filled with neutrophils, macrophages, red blood cells, and proteinaceous edema fluid. AEI cells slough off of the basement membrane and are replaced by hyaline membranes. The remaining AEII cells cover the denuded basement membrane to ultimately assume the phenotype of AEI cells [7].

The pulmonary vasculature also suffers damage in the course of ALI/ARDS, and some surmise that the degree of pulmonary endothelial activation and injury may in fact be an important determinant of clinical outcomes [8]. In the normal lung, vascular endothelium is a dynamic border between circulating blood and lung tissue. It is a monolayer that acts as a non-adhesive surface for platelets and leukocytes, producing a variety of regulatory factors such as nitric oxide (NO) and prostaglandins. This endothelial monolayer typically turns over very slowly; in healthy subjects, only 1 to 3 vessel wall-derived endothelial cells per cc blood can be identified in the circulation [9]. In situations of injury, however, the endothelium loses its antithrombotic property, and the number of circulating endothelial cells is much higher. In ALI/ARDS, cells lining the endothelium become swollen and vacuolated, with gaps present between adjacent endothelial cells where there are normally tight junctions [1]. These cells become activated, allowing attachment and diapedesis of inflammatory cells from the bloodstream across the endothelium. To repair damaged endothelium, local mature endothelial cells adjacent to the area of injury may be important; however, these mature cells have a very limited proliferative capacity. In the past decade, circulating bone marrow-derived cells similar to embryonal angioblasts have been identified [10]. These cells have the potential to proliferate and differentiate into mature endothelial cells [9;10].

Pulmonary hypertension (PHTN) is another associated feature of ALI/ARDS, resulting from hypoxic and mediator-induced vasoconstriction, thromboembolism, or local compression by edema fluid and fibrosis [11]. The use of vasodilators to decrease pulmonary arterial pressure has been associated with short-term improvements in refractory hypoxia [12] and may temporarily improve a patient's condition until he or she can respond to other therapeutic agents to reverse the underlying cause for ALI/ARDS (e.g. antibiotics). In ALI/ARDS, pulmonary hypertension develops in a short period of time, so it is less likely that long-term, irreversible lesions of PHTN (e.g. vascular wall remodeling) are present to such a significant extent that PHTN would be refractory to vasodilatory therapy. Therefore, therapy directed at PHTN related to ALI/ARDS along with conventional treatment of this disorder may benefit these patients.

A notable feature of the pathophysiology described in ALI/ARDS is the acuity with which this damage occurs, usually in a matter of hours. This contrasts to other pulmonary disorders such as emphysema or cystic fibrosis, where there is more indolent damage to the alveolar epithelium, or primary pulmonary hypertension, where there is progressive remodeling of the pulmonary vasculature resulting in symptomatic disease after years have transpired [7]. Given the dramatic, fulminant nature of ALI/ARDS and the lack of available therapies, investigators and clinicians alike are eager to explore new therapies for these disorders [13]. As a result, the role of stem cells in the repair of the damaged lung is receiving increasing attention (1). Like other areas in stem cell biology, investigations regarding the contributions of stem cells in repair of the lung have been difficult, although slow progress is being made. Problems facing stem cell therapy for ALI/ARDS are not unique to this disease process, but are present for all illnesses affecting the lungs and in other organ systems as well [15]. Many issues need to be addressed; including what processes in this syndrome should be treated: epithelial damage, endothelial damage, or both? What point in the disease process should treatment be given? How should therapies be administered? By what mechanisms do stem cells mitigate disease, and are animal models reflective of what occurs in human disease? As cardiologists have forged forwards in stem cell therapy for diseases of the heart and clinical trials have been published [16-18], the answer to many of these questions is not forthcoming.

Repair of damaged alveolar epithelium and endothelium likely correlates with better outcomes in ALI/ARDS. This review will focus on the role of stem cells in the repair of lung injury, specifically those studies that focus on the regeneration of lung epithelium and endothelium. Potential therapies for pulmonary hypertension will be highlighted as well. Investigations that have been performed in animal models will be described, and how these might apply to human disease. For purposes of clarity, the use of the acronym ALI will refer to ALI and ARDS in this review, since these are pathophysiologically similar (Table 1) [19].

STEM CELLS IN PULMONARY BIOLOGY

Much of the difficulty with the study of stem cells in lung diseases has been the generalized lack of agreement over what defines a "stem cell" [14]. Suggested defining criteria for a stem cell are: the *capacity for limitless self-renewal, no features of a differentiated cell,* and *multipotency,* referring to the capacity of a stem cell's progeny to form at least one, and often several lineages of differentiated cells [7;20]. This contrasts with pluripotency, where a cell is able to give rise to every other type of cell in the body, a feature of embryonic stem cells. Of note, some scientists argue that pluripotency should be a requirement for a cell to be called a "stem cell". Additionally, some feel that a putative stem cell should reside in a discrete microenvironment or "niche" [7]. Another defining characteristic of a stem cell is its relative quiescence, with infrequent proliferation in the steady state that apparently serves to protect it from injury via mutagenesis [21]. To compound this difficulty with the precise definition of a stem cell, there is also the added anatomic and functional complexity of the lung, an organ comprised of approximately 40 morphologically different lineages. These issues can partially explain the slow progress for investigations in stem cell biology within the pulmonary arena. The lack of consensus over what constitutes "stemness" will likely persist for years to come. However, research has progressed in identifying some of the operative mechanisms in lung repair where cells that have at least *some* of these properties may be important. These cells are most precisely termed "multipotent stem cells", and possess the characteristic ability to differentiate into a limited range of cell lineages within the lung [7;20]. They have been identified both in the lung and in the bone marrow.

Table 1. American-European Concensus Definitions for the Acute Respiratory Distress Syndrome (ARDS) and Acute Lung Injury (ALI) [2]

Acute lung injury (ALI): a syndrome of acute and persistent lung inflammation with increased vascular permeability, characterized by

- ❖ Bilateral radiographic infiltrates.
- ❖ A ratio of the partial pressure of arterial oxygen to the fraction of inspired oxygen (PaO_2/FiO_2) between 201 and 300 mmHg, regardless of the level of positive end-expiratory pressure (PEEP). The PaO_2 is measured in mmHg and the FiO_2 is expressed as a decimal between 0.21 and 1.00.
- ❖ No clinical evidence for an elevated left atrial pressure. If measured, the pulmonary capillary wedge pressure is 18 mmHg or less.

Acute respiratory distress syndrome (ARDS): definition the same as ALI except that the hypoxia is worse, requiring

- ❖ PaO_2/FiO_2 ratio of 200 mmHg or less, regardless of the level of PEEP

WHERE DO CELLS THAT REPAIR THE LUNG ORIGINATE? (TABLE 2)

Cells Intrinsic to the Lung

Multipotent stem cells within the lung are proposed to exist in a specialized environment comprised of a subset of cells and cellular components, known as a "niche". The niche participates in environmental signaling with factors residing in or on stem cells, and may contribute to these cells' "stemness". Cells within a niche go through cell cycle very slowly but have the potential to become a myriad of cell types [20;22]. Stem cells within niches are capable of differentiating in the setting of injury. If such niches existed within alveoli, they might be important in repair of ALI given the degree of alveolar epithelial injury present [22].

However, no definitive intra-alveolar stem cell niche has been identified, although the existence of niches in the proximal and distal conducting airways has more widespread acceptance [20].

Sources of intrinsic stem cells near the alveolus that may function in alveolar epithelial repair are cells of the bronchoalveolar duct junction (BADJ). These cells are strategically located, and appear to be progenitors of both Clara cells and alveolar cells *in vitro* [23]. Bronchoalveolar stem cells (BASCs) are a type of Clara cell resistant to naphthalene injury, located at the intersection between conducting and respiratory epithelium. They express surfactant protein C, and the Clara cell-specific marker CCA.

Table 2. Progenitor cells with a potential role in the repair of ALI

	Markers	Refs
Cells intrinsic to lung		
Niche cells	Unknown	[3;4]
Bronchoalveolar Duct Junction Cells	CCA+, SP-C+	[5]
Alveolar Epithelial Type II Cells	E-cadherin-, surfactant proteins, TTF-1	[6-8]
Lung Side Population Cells	CD45+/CD45-, Sca-1+	[9]
Cells extrinsic to lung		
Bone Marrow-derived Cells and Circulating Progenitor Cells (subtypes below)	Linage -	[10]
Immature cells	Sca-1+	[11]
Mesenchymal stem cells	CD45-/CD11-/CD34-	[12;13]
Multipotent adult progenitor cells	CD34-/ CD44-/ CD45-/c-kit-/ MHC-I-/MHC-II-/CD13+/SSEA-1+	[14]
Endothelial progenitor cells	CD34+/CD133+/Flk-1+/Tie-2+/KDR+/VE-cadherin+/vWF+	[15;16]

CCA=Clara cell antigen, SP=surfactant protein, SSEA-1=stage-specific antigen-1, MHC=major histocompatibility class, Flk-1=receptor for vascular endothelium-derived growth factor, KDR=kinase domain repeat, vWF=von Willebrand factor, TTF=thyroid transcription factor.

The best-studied and most widely accepted precursor for cells of the alveolar epithelium are AEII cells, characterized by lamellar bodies, apical microvilli, and a cuboidal shape. These cells have been regarded as the local stem cell responsible for the repair of the alveolar epithelium based on studies almost 40 years old [24]. Notably, AEII cells do not follow the typical stem cell paradigm in that they function as differentiated cells in the uninjured lung, and are considered to be unipotent (able to differentiate into one cell type), rather than multipotent. Nevertheless, AEII cells have been observed to re-enter the cell cycle to ultimately become AEI cells after lung injury [25]. During the course of lung repair, it is unknown if differentiation from AEII to AEI cell is mediated by cell division, but it has been demonstrated that in certain culture conditions, AEII cells can alter their morphology to become AEI cells (i.e. lose lamellar bodies) without cell division, and then in different culture conditions, regain their AEII characteristics [26].

Another potential stem cell type intrinsic to the lung, termed lung side population (SP) cells, has mesenchymal, epithelial, and even endothelial potential. Their site of residence within the lung is as of yet unidentified. They represent 0.03-0.07% of all lung cells [27], but have been identified in other body tissues as well. SP cells in lung may in fact be marrow-derived, and take up residence in the lung secondarily [28]. These cells are identified by characteristics on fluorescent-activated cell sorting (FACS). When stained with the vital DNA dye, Hoechst 33342, and excited with UV laser, these cells fluoresce with a characteristic low blue and low red pattern, related to efflux of the Hoechst dye due to a transporter on the SP cells. SP cells are lineage-negative, and are characterized as enriched primitive hematopoietic precursors with multipotent stem cell abilities. Approximately 75% of lung-derived SP cells express CD45, while bone marrow-derived SP cells express this marker universally [27;29]. SP cells from the lung may have a role in homeostasis and repair of vascular endothelium, where interaction of smooth muscle and endothelial cell types is known to be important. To support this, investigators have demonstrated the co-localization of CD31, an endothelial marker, on CD45 SP cells, suggesting that lung SP cells can play a role in homeostasis of the pulmonary vasculature [27].

CELLS EXTRINSIC TO THE LUNG

In the past decade, there has been increasing interest in the prospect of bone marrow-derived cells converting to lung cells in situations of lung injury. This phenomenon was first reported by Krause and colleagues in 2001, who were able to demonstrate that transplantation of a single, lineage-depleted male mouse bone marrow cell into irradiated female recipients resulted in engraftment in the alveolar epithelium (including AEII alveolar epithelial cells) as well as the bronchial epithelium [30]. Subsequently, numerous investigations in murine transplant models have demonstrated conversion of a percentage of recipient's lung cells into the donor's genotype, and these cells have been noted to exhibit phenotypes of alveolar epithelium and endothelium [31-33].

In recent years, a variety of cell sorting and isolation procedures on bone marrow cells prior to their use in transplant have developed. These techniques can specifically select for multipotent precursor cells in order to study their role in lung injury. One straightforward approach is to isolate the most immature cells within the bone marrow or circulation, with the expectation that these can differentiate into numerous lung cell types (both endothelial and epithelial). Investigators have isolated such cells within a fraction of peripheral blood mononuclear cells using magnetic beads that select for the Sca-1 antigen, a very primitive stem cell marker [34]. Another technique is to plate bone marrow cells on plastic, and to subsequently identify different markers on these cells' surface in order to subclassify them. In this type of system, cells adherent to plastic that possess the marker CD31 are consistent with endothelial progenitors, while those that express CD45 are of myeloid lineage, and those that synthesize matrix proteins represent fibroblasts. Non-adherent cells in culture represent hematopoietic stem cells [35;36]. Methods have also been developed to isolate presumed mesenchymal stem cells (MSCs) that are able to differentiate into a variety of mesodermal types such as bone, adipose tissue, and stromal cells [37]. Whole bone marrow is cultured for several days, any non-adherent cells are discarded, and the remainder successively immunodepleted with anti-CD11 antibodies (to eliminate macrophages), along with anti-CD34 and anti-CD45 antibodies (to eliminate hematopoietic cells) [32;38]. Other groups have isolated rare cells that co-purify with mesenchymal stem cells, termed multipotent adult progenitor cells (MAPCs) [39]. These cells are derived by culturing bone marrow mononuclear cells from mice on fibronectin, laminin, type IV collagen, and matrigel, and then maintaining them in expansion media. Cultured cells are collected and immunodepleted with anti-CD45 and anti-Ter119 antibodies, and selected cells (about 20% of all cells) maintained in culture. Isolated cells express low levels of Flk-1 and Sca-1, and are MHC-class-I and CD44 negative, similar to what is observed in human MAPC [40;41]. These cells can be extensively expanded through many population doublings, and as their name implies, can differentiate into a variety of cell lineages. In order to isolate and study endothelial progenitors from whole bone marrow, peripheral blood, or cord blood, magnetic microbeads that positively select for the endothelial markers CD133 or CD34 have been employed [42]. Alternatively, different adhesion culture methods, such as culturing peripheral blood mononuclear cells on fibronectin [43], have also been used to isolate endothelial precursors. Depending on the type of repair that is being studied (e.g. alveolar epithelial repair) a bone marrow-derived cell with certain markers or capabilities might be chosen.

INVESTIGATIONS DEMONSTRATING ALVEOLAR EPITHELIAL REPAIR

With Intrinsic Lung Stem Cells

As previously mentioned, some investigators question the true "stemness" of the AEII cell given its possession of markers of cell maturity and its ability to function in a differentiated role. Others believe that there may be only a subpopulation of AEII cells that can truly perform this role [21].Reddy and colleagues [44] observed four distinct subpopulations of type II cells in a rat model after subjecting animals to hypoxia: (1)non-proliferative AEII cells that appeared undamaged, (2)proliferative AEII cells that appeared undamaged, (3)non-proliferative AEII cells that were damage-susceptible, and (4)proliferative AEII cells that were damage-susceptible. The investigators hypothesized that the lungs of animals injured by hypoxia would contain a transiently amplifying population of cells that was both proliferative and resistant to damage (type 2 above) that would in fact represent the AEII progenitor responsible for repair of the alveolar epithelium. In contrast to a "true" stem cell as previously defined, these stem cells would be more differentiated in function and have a more finite lifespan, but still would be able to function in lung repair. Alternatively, the authors postulated that the "true" alveolar epithelial stem cell might in fact reside within the nonproliferating, undamaged population (type 1 above). After examining a variety of markers important in cell-cell interactions and nucleus-to-cell surface interactions, they discovered that AEII cells subjected to hypoxic injury variably expressed E-cadherin by immunohistochemical staining, fluorescent-activated cell sorting, and separation with antibody-coated magnetic beads. The population of AEII cells that did not express E-cadherin was significantly more damage resistant, more proliferative, and had a high level of telomerase activity, all characteristics consistent with a transiently amplifying population of AEII cells involved in repairing damaged epithelium. These findings support the longstanding theory that AEII cells are progenitors of AEI cells, but stipulate that not every AEII cell may function in this role.

Kim and colleagues [23] have demonstrated that cells from the BADJ, termed bronchoalveolar stem cells (BASCs) may in fact be precursors for AEI cells, or instead, precursors for AEI cells via an AEII cell intermediate. These investigators subjected mice to inhaled bleomycin to induce alveolar epithelial injury, whereupon the number of BASCs at the bronchoalveolar duct junction increased significantly out to 14 days. They speculated that these cells could ultimately differentiate into AEI cells, but did not demonstrate that this in fact would occur. Investigation of BASCs and their role in ALI remains a topic of active investigation.

Summer and colleagues examined the expression of epithelial and mesenchymal genes in SP cells, and noted these cells' ability to variably express alpha-smooth muscle actin and keratins, properties of the epithelial cells in the lung. The function of SP cells in alveolar epithelial repair is still largely unexplored [28].

With Cells of Extrinsic to the Lung

Krause and colleagues [30] provided initial evidence that a single, lineage-depleted bone marrow cell transplanted into lethally irradiated recipients would lead to engraftment into the recipients' bronchi and alveolar epithelium, as well as other organs of the animal. Interestingly, the degree of engraftment was most robust in alveolar epithelium (up to 20% in some animals at 11 months) compared to the epithelium of other organs, possibly the result of radiation-induced pneumonitis in the transplant recipients. This underscores the importance of damage to an organ eliciting its repair. In donor cells that apparently engrafted, the expression of mRNA for surfactant proteins, an AEII-cell specific feature, was also noted to be present. Thus, it appeared that the donor cells were behaving in a tissue-specific fashion.

In another investigation, when Jiang and Jahagirdar [39] isolated MAPCs from ROSA26 mice and then transplanted them into non-obese diabetic/severe combined immunodeficiency (NOD/SCID) mice, approximately 2-10% engraftment of the lung was observed after intravenous infusion of 10^6 MAPCs. In this investigation, not all animals had been subjected to radiation injury. This difference could explain the lesser degree of engraftment than that reported by Krause. In another model utilizing parabiotic mice, green fluorescent protein-transgenic animals were joined to wild-type littermates [33]. The wild-type mice were then subjected to lung injury with radiation, intratracheal elastase, or both. In the alveoli of these animals, Abe and colleagues were able to demonstrate the presence of cells doubly positive for GFP and cytokeratin that they termed "alveolar epithelial type I-like". They did not observe cells with features of AEII cells, nor did they quantify the degree of engraftment present in the animals' lungs. All three of these studies provide evidence that circulating cells may function in repair of injured alveolar epithelium, including those that originate in the bone marrow.

Kotton and colleagues utilized the technique of plating bone marrow on plastic to help identify which bone marrow cells would engraft in the lungs of injured mice [45]. These investigators obtained whole bone marrow from mice constitutively expressing the lacZ/neomycin resistance transgene. Upon engraftment of these donor cells into the recipient animals' tissues, they could be identified by the use of X-gal staining that would mark those cells with lacZ expression. The collected cells were plated on plastic, with adherent cells harvested at a week. Recipient mice were subjected to lung injury with intratracheal bleomycin, and then given approximately 1-2 million of these cells from the total cell fraction intravenously. When the recipient animals' lungs were examined at time points from 5 to 30 days after cell infusion, cells with an AEI morphology that were of donor origin were determined to be present that stained doubly positive for X-gal and T1α, an AEI cell marker . These investigators could not demonstrate lacZ positive AEII cells, but were able to demonstrate donor cells within the pulmonary circulation at early time points after donor cell infusion (i.e. 1 and 2.5 days). Interestingly, it was determined that only 10% of the cells initially infused were positive for T1α by immunohistochemical staining. Also important was the observation that lung engraftment was the most robust in animals after bleomycin injury (4/4 animals), but was infrequent in those not injured (2/9) (p=0.02).

Ortiz and colleagues [32] transplanted MSCs (from males) into female C57Bl/6 mice, and observed engraftment in the lungs that was most significant in animals subjected to lung injury with bleomycin. When AEII cells were isolated from these transplanted animals, male DNA measured by RT-PCR was 3-fold higher in the animals that were bleomycin-injured. Most clinically relevant, administering MSCs after bleomycin challenge reduced inflammation, decreased the degree of collagen deposition, and decreased MMP-2 and MMP-9 transcription within the lungs of animals when MSCs were given immediately after bleomycin injury. Rojas and colleagues [38] transplanted similarly prepared MSCs derived from bone marrow of male mice expressing a green fluorescent protein (GFP) transgene into animals that had been treated with intratracheal bleomycin. A subgroup of these recipients had also received of busulfan, a myelosuppressive agent. The investigators were able to demonstrate co-localization of GFP protein (from the donor animal) and cell-type specific markers for AEI and AEII cells 14 days after transplantation. Interestingly, the number of donor-derived cells was significantly greater in lungs of transplanted animals that had been given busulfan prior to the administration of bleomycin; animals that had received both bleomycin and busulfan but were not given MSCs had a much higher mortality than those transplanted with MSCs. This suggests that both exogenously administered MSCs as well as cells from the animals' own marrow are important in engraftment and repair of the acutely injured lung.

In another study by Yamada and colleagues [34], investigators subjected C57Bl/6 mice to bone marrow suppressive doses of radiation, after which they administered Sca-1+ bone marrow-derived cells from GFP transgenic mice, and then allowed the animals to reconstitute their bone marrow. Three weeks later, the animals were subjected to inhaled lipopolysaccharide (LPS). Utilizing immunofluorescent staining to locate the GFP-positive cells from the donor, along with cytokeratin to mark AEI cells, the co-localization of GFP and cytokeratin was observed to a greater extent in the LPS-injured animals. When the investigators performed these experiments in a different sequence, administering LPS after radiation, then immediately giving the animals an infusion of bone marrow-derived Sca-1+ cells, the results were similar. Interestingly, animals given insufflated LPS after radiation but not Sca-1 + cells developed emphysematous-like lesions. Similar to the studies by Ortiz and Rojas, an infusion of stem cells *after injury* was associated with the most prominent engraftment.

INVESTIGATIONS DEMONSTRATING ENDOTHELIAL REPAIR

With Cells Intrinsic to the Lung

SP cells intrinsic to the lung may have a role in the repair of damaged endothelium observed in ALI. Summer and colleagues isolated SP cells from the lungs of embryonic mice at day 17.5, a time of growth for vascular elements within the mouse lung[46]. It was determined that these cells constituted approximately 1% of the total lung population in these embryonic animals, a much higher percentage than that observed in the adult mouse. In the SP CD45 negative subset, genes associated with an endothelial phenotype were highly represented, including vWF, kinase insert domain containing receptor (KDR), Tie1, and vascular-derived growth factor (VEGF)-c. Utilizing this gene profiling data, two populations of SP CD45 negative cells were identified, namely those that were vWF positive/smooth muscle actin (Sma) positive cells, found in the muscular layer of large vessels, as well as vWF positive/intercellular adhesion molecule (ICAM)-2 positive cells within the endothelial layer of small vessels. These CD45 negative SP cells were further characterized based on expression of CD31, then cultured in media specific for smooth muscle growth or for endothelial growth. SP cells plated in smooth muscle growth media that were CD45-/CD31+ displayed characteristics of mature smooth muscle; however, CD45-/CD31- SP cells cultured in endothelial growth media failed to differentiate. When CD45-/CD31+ and CD45-/CD31- SP cells were co-cultured in endothelial growth media, however, vascular tube-like networks formed that stained positively for von Willebrand factor (vWF), suggestive of endothelium formation. These findings demonstrate the ability of SP cells to form cells of the endothelium under appropriate conditions *in vitro*; whether they actually are capable of these activities *in vivo* remains under active investigation.

With Cells Extrinsic to the Lung

In the past few years, several reports of bone marrow cells' ability to assume properties of endothelium have been published. These studies are relevant to ALI repair given the extent of endothelial injury present in this disorder. Jiang and colleagues [39] attempted to induce differentiation of MAPCs into endothelial cells to prove their ability to form cells of the mesodermal lineage. When these cells were examined at a baseline they lacked endothelial markers CD31 and vWF, but did express low levels of Flk-1, as previously mentioned. However, when these murine MAPCs were cultured on fibronectin-coated plates and supplemented with VEGF for 14 days, a majority of them (~90%) assumed an endothelial phenotype, with significant expression of CD31, vWF and Flk-1, suggesting that certain bone marrow cells could form cells of the endothelium under appropriate conditions.

In the model of LPS-induced lung injury [34], Yamada and colleagues demonstrated that injury alone would mobilize progenitor cells with endothelial features into the circulation. When C57Bl/6 mice were given intranasal LPS, the number of Sca-1 and Flk-1 (an endothelial marker) positive cells in the circulation significantly increased. When PBMCs from these animals subjected to LPS were cultured for 7 days on fibronectin, they also expressed the endothelial markers Flk-1 and PECAM-1 with greater intensity than did controls. Finally, in a more functional assay, sublethally irradiated animals given intranasal LPS followed immediately by a bone marrow cell infusion from GFP-transgenic animals had observable GFP+/CD34+/CD45- cells within their lung parenchyma. The studies by Jiang and Yamada indicate the potential of cells derived from bone marrow to assume properties of vascular endothelium. Further isolation and manipulation of these cells may be necessary to fully understand their role in repair of damaged lung endothelium.

THE ROLE OF INJURY IN STEM CELL ENGRAFTMENT

Eliciting lung injury in animal models has allowed investigators to explore the role of both intrinsic and extrinsic stem cells in the repair of lung injury. One recurrent theme highlighted by the investigations described is the requirement for *lung injury* prior to detectable engraftment and repair by multipotent stem cells [47]. Another recurrent theme is evidence of cross-talk between the bone marrow and the *injured* lung. In animals with minimal lung injury that are transplanted, investigations have demonstrated virtually no engraftment [32]. Complementary to this, animals subjected to lung injury without an intact, functioning bone marrow exhibit inadequate lung repair [32;38], underscoring the potential importance of bone-marrow derived stem cells in repair. However, it also remains possible that lung injury induces intrinsic lung stem cells residing in local niches to participate in damage repair [33].

HOW CAN STEM CELLS BE USED THERAPEUTICALLY IN ALI?

Mobilization of Endogenous Stem Cells without Transplantation

Given the intricacies of purification and administration of exogenous stem cells in the therapy of disorders such as ALI, mobilization of endogenous stem cells within those who are ill is an attractive alternative in manipulating stem cells to participate in repair. This is particularly relevant in ALI given its acuity and high short-term mortality. A treatment given soon after diagnosis of this disorder is more likely to be efficacious than one given many days after diagnosis, when irreparable damage might have occurred.

Human stem cell mobilization with the selection of immature cells for transplant has become preferable in clinical transplantation because of the higher yield of immature cells, the shorter time frame to reach successful repopulation, the reduced need for interventional procedures, and pain advantages compared to harvesting bone marrow cells through traditional bone marrow aspiration [48]. The growth factors granulocyte colony-stimulating factor (GCSF) and granulocyte-monocyte colony stimulating factor (GMCSF) are two medications widely used to mobilize blood precursors in the fields of oncology and transplantation. GCSF increases the expression and function of adhesion receptors on the surface of hematopoietic progenitors and stromal cells, and may influence marrow cell homing [49]. It is the most established agent for the mobilization of stem cells in current clinical practice [50]. GCSF acts by inducing stem cell proliferation within bone marrow, and subsequently causes these cells' release into the peripheral blood in part by causing neutrophils to secrete proteinases (e.g. elastase). These proteinases cleave adhesive bonds between stromal cells and hematopoietic progenitor cells, and cleave the cytokine stromal cell derived factor-1 (SDF-1), from its receptor, CXCR-4, found on stromal and progenitor cells. High levels of SDF-1 induced through this cleavage appear to cause in CXCR-4 positive cells to exit the bone marrow into the circulation [51]. Another important role of SDF-1 may be the induction of progenitor (specifically CD34 positive progenitors) cell migration, as this cytokine was shown to be active in migration of progenitors both *in vitro* and *in vivo* [52]. Similar to GCSF and GMCSF, vascular endothelium-derived growth factor (VEGF) has also been demonstrated to stimulate release of endothelial progenitors from the bone marrow, and levels of VEGF in the bloodstream are increased in situations of injury such as limb ischemia, or status post coronary artery bypass surgery [9]. Erythropoietin (Epo) is another cytokine that might prove useful in the mobilization of progenitor cells, and has been shown to increase the mobilization of endothelial progenitors in both animal models [53] as well as the circulation of humans [54]. In general, the more potent an agent is in mobilizing progenitor cells, the more inflammatory it is, which poses concerns for administering such agents to patients with ALI who are in a highly inflammatory state. GMCSF and GCSF are two of the more inflammatory agents, known to increase inflammation with the enhancement of leukocytosis, while Epo and VEGF are less so. Further studies are necessary to determine a role for exogenously administered growth factors in patients with ALI to stimulate stem cell release. If these agents prove useful in improving outcomes, they could serve to enhance or augment available therapies for this disorder.

Stem Cells as a Vector to Deliver Gene Therapy for Secondary Pulmonary HTN in ALI

Investigators have explored different methods to deliver vasodilatory agents in models of pulmonary hypertension (PHTN) through the use of stem cells as vectors for gene products, the theory being that the gene product will be expressed once the cell is integrated in the lung. This could help solve the problems with utilizing viruses as vectors that included inciting an unwanted inflammatory response, and underwhelming transgene expression *in vivo* [55]. Stem cells of the pulmonary endothelium theoretically could be useful vectors for vasodilatory genes if they incorporate into the endothelium, and therefore might be therapeutic for patients with ALI who oftentimes have severe PHTN. Nagaya and colleagues [56] cultured cord blood cells on fibronectin for 8 days, demonstrating that a majority of adherent cells would stain positively for DiI-acLDL and FITC-labeled lectin, and express the endothelial-specific antigens KDR, VE-cadherin, and CD31. The authors therefore assumed that most adherent cells were endothelial progenitors. The cells were then cultured with a GFP-gelatin complex containing plasmid DNA for adrenomedullin, a vasodilator peptide. The endothelial progenitors rapidly phagocytosed this material. An intravenous infusion of these cells was then administered to nude rats in a model of monocrotaline (MCT)-induced PHTN, and these transplanted endothelial progenitors appeared to incorporate into the pulmonary vasculature. Both mean pulmonary artery pressure and pulmonary vascular resistance was significantly decreased in animals that had been given the endothelial progenitors, even in a group of animals that had been given endothelial progenitors that had ingested a sham plasmid. However, animal survival was greatest in the group that had received the progenitor cell infusion with the DNA plasmid. Zhao et al, in a similar protocol of MCT-induced PHTN in rats, attempted to reverse this condition through the use of "endothelial-like progenitor cells" (ELPCs) isolated from rat bone marrow [57]. They isolated mononuclear cells from the bone marrow of Fisher-344 rats, resuspended these cells in endothelial growth media, and finally plated them on gelatin-coated tissue culture flasks for 7-10 days to isolate the ELPCs. The cells thus isolated were then transduced with either human eNOS within a plasmid vector, or with an empty vector as control. Twenty-one days after MCT had been administered to animals, baseline right ventricular systolic pressures (RVSPs) were measured, and the animals received intravenous saline, ELPCs without eNOS DNA, or ELPCs with eNOS DNA. Fourteen days later, RVSPs were measured again. The administration of the ELPCs with the eNOS DNA significantly reduced the RVSP from pretreatment levels in animals that had been treated in this fashion. Also, treatment of the animals with ELPCs containing a null vector had stabilization of their RVSPs, but no progression to higher pressures, in contrast to what the saline-treated group exhibited. When the lungs of animals were microscopically examined, the animals that had received eNOS-transduced ELPCs had a more normal appearance of their pulmonary circulation, with microvascular perfusion (as measured by the perfusion index) on par with what was observed in non-MCT treated animals. Kanki-Horimoto and colleagues performed nearly identical experiments as this group, save for their use of MSCs as the cell of transplant [58]. These were isolated from Sprague-Dawley rats and cultured in minimal essential medium, where they became confluent. They were then passaged 3-5 times, and were then transfected with DNA for eNOS via an adenovirus. Animals were treated with MCT to induce PHTN, and 21 days later were infused with saline, MSCs with the eNOS vector, or MSCs without eNOS vector. As was seen in the prior study, RVSPs were significantly lower in the groups that had been treated with MSCs with or without the eNOS vector; those that had received the vector had pressures on par with non-monocrotaline treated controls. Survival time was also significantly longer in those groups receiving MSCs (with or without vector) compared to the group that had received saline alone.

Important features of the aforementioned studies are the fact that physiologic benefit was seen in a short period of time, and survival benefits were also demonstrable in the short run. More intriguing is the fact that benefits were seen from infusion of the stem cells alone without the addition of the vasodilatory product. Development of this type of therapy for use in ALI patients with secondary PHTN might prove useful in that it "buys time" for correction of the underlying illness (i.e. sepsis). Also, one might speculate that mere integration of the stem cells into damaged pulmonary vascular endothelium may be of benefit in terms of providing a scaffold for repair of the lung to occur. Alternatively, these cells may have a paracrine effect important in the recruitment of other cells involved in repair.

Stem Cell Infusions to Repair Lung Injury in ALI

While clinical trials of stem cell therapy for cardiovascular diseases progress [16;17;18], such trials are likely in the distant future for disorders such as ALI. Importantly, studies published regarding the use of stem cells for cardiovascular diseases have revealed either no clinically significant benefit [18], or minimal benefit to patients [16;17]. It is worth noting, however, that none of these studies have been associated with adverse outcomes up to one year after the therapy was given. Improved study designs with cells that are more highly targeted to repair a given cell type may ultimately prove to be more beneficial. Technology that has been used in patients with cardiovascular disease, specifically intracoronary injection of progenitor cells, could readily be modified to deliver cells to patients with ALI (i.e. through an intravascular route). Critically ill patients with no other therapeutic alternatives, such as those with ALI, may stand to benefit the most from cell-based therapies.

WHAT ARE THE PERSISTENT CONTROVERSIES IN STEM CELL RESEARCH AS IT PERTAINS TO LUNG REPAIR?

Design of Animal Models of ALI

In animals, a variety of methods can be used to provoke lung injury in order to mimic human ALI. Whether or not these methods accurately re-create this disorder in humans is debatable. One common method to study stem cell engraftment in animal models illustrated in the studies discussed is to administer total body irradiation (TBI), followed by administration of a lung toxicant.

Total body irradiation (TBI) has been used as a preparative regimen prior to stem cell transplantation since the earliest days of this type of therapy. TBI facilitates marrow chimerism, where the recipient's bone marrow cells are replaced by donor bone marrow cells [59]. AEI cells are exquisitely sensitive to the effects of radiation, whereas AEII cells are more radiation-resistant. Maximal doses of TBI in humans approximate 15 Gy, with higher doses producing primarily lung toxicity. However, lower doses of radiation may result in lung damage. In a mouse model, only 1.2 Gy of TBI resulted in lung hypocellularity in mice, with pulmonary capillary breakdown and extravasation of erythrocytes into the airspaces [31]. Radiation-induced lung injury resembles ALI/ARDS through its damage to the alveolar epithelium; radiation pneumonitis (*without* concomitant administration of a pulmotoxic agent) was formerly used as a model for ALI/ARDS [60]. Radiation given to humans affects the lung in a dose-dependent manner [61]. Therefore, it is not surprising that the degree of engraftment correlates directly with the dose of radiation in animals [59]. The location of radiation administration may also factor in to the degree of engraftment, with greater engraftment observable near the site of radiation [47]. These studies suggest that careful attention needs to be given to the radiation regimen planned in animal studies of stem cell transplantation in order to take into account the degree of lung damage that may occur as a result of TBI or other radiation administration.

The type of substance used to chemically elicit lung injury should also be carefully evaluated when performing animal studies. Intratracheal bleomycin is commonly administered for these purposes. Its exact mechanism of action is unknown, but AEI cells are very sensitive to this type of injury. Bleomycin injury also results in increased cellularity and fibrosis within the lung related to the direct effect of this drug on fibroblasts that initiates a reparative response. Inhaled doses of approximately 4U bleomycin/kg animal are typically used to produce injury [32;38;45]. Inhaled lipopolysaccharide (LPS), also known as endotoxin, can also be used to induce lung injury at doses of approximately 20 mcg/animal [34]. LPS activates leukocytes and leads to an increase in alveolar-capillary permeability, critical features of ALI/ARDS. These factors all contribute to the development of pulmonary edema [62]. Another inhaled pulmotoxic agent bears mentioning: naphthalene. This agent is utilized in mice models to elicit Clara cell injury in studies where these cells are to be specifically examined [63].

Some investigators have utilized both radiation and chemical agents in series to induce lung damage. Aliotta and colleagues utilized increasing doses of TBI in conjunction with injection of intramuscular cobra venom [59], while Abe and colleagues utilized a combination of TBI and intratracheal elastase [33]. Each of these investigations demonstrated an additive effect on engraftment of donor cells when animals were subjected to both chemical and radiation injury. However, even though the use of radiation and chemical agents is the most common way to elicit ALI in animal models, it still may not be optimal. Beckett and colleagues recently demonstrated that leukocytes, not AEI or AEII cells, represented the majority of donor-derived cells within the lungs of irradiated (TBI, 10Gy) C57Bl/6 mice that had been transplanted with resuspended total bone marrow cells [64]. A much smaller percentage (0.2-0.3%) had markers consistent with AEII cells. Further lung injury to these recipient animals with either endotoxin or nitric oxide did not enhance the number of alveolar epithelial cells that engrafted. In sum, injuries to the lung produced via radiation and/or chemical means should be carefully planned to account for the type and extent of lung injury produced in an animal model. Also, the development of newer animal injury models displaying more robust features of engraftment would be useful in further delineating the mechanisms required for clinically significant transplantation.

Identification of Donor Cells

Techniques used to identify engrafted stem cells within the lungs of transplanted recipients have oftentimes been plagued by inconsistencies and uncertainty over the validity of the methods employed. As a result, consensus has not yet been reached over whether or not donor cells truly engraft into the lungs of recipients, or whether a fusion phenomenon is being observed [22]. Immunostaining and fluorescent microscopy has been heavily relied upon to identify the presence of donor cells within a recipient's lung, and has been integral to advances in lung stem cell research. However, these techniques are well-known to be prone to artifact. Antibodies used in immunohistochemical staining react with other cells than their intended targets (false positives) [65]. The lung has high background immunofluorescence due to such factors as fixative exposure and its elastin content [66]. Further, microscopy is a rather subjective technique [65]. To compound these difficulties, given the nature of the lung's architecture, with its capillary-containing septae in the alveoli, immunostaining of blood cells can be misinterpreted as "engraftment". Recent work by Kotton and colleagues [66] attempted to sort through the validity of earlier findings that used immunostaining and microscopy to document engraftment, including some of their own previous work. In this study, they performed analysis based on histology, fluorescent-activated cell sorting, molecular biology, and a lineage-specific reporting system to exclude background events that could be misinterpreted as engraftment (i.e. non-specifically labeled cells, autofluorescence).

These investigators transplanted GFP+ whole marrow or hematopoietic stem cells from mice transgenic for surfactant protein C-GFP reporter into recipients that had been irradiated, some of whom had also been subjected to lung injury with bleomycin. If the donor cells were functioning as AEII cells in recipients, they should fluoresce in recipient animals. When background events were excluded, bone-marrow derived cells did not appear to contribute to the AEII population in the recipient animals.

Finally, part of the difficulty in donor cell identification must be attributed to the low prevalence of these cells in recipients. This is illustrated in studies of SP cells, an extremely rare (<1%) subtype of cells within lung. The low number of these cells and others used in transplant necessitates strict adherence to all aspects of an isolation protocol to ensure accurate isolation and identification of these types of cells [29].

ARE BONE MARROW CELLS IMPORTANT IN IMPROVING OUTCOMES FOR ALI IN HUMANS?

Examination of the human lung for potential engraftment by bone marrow derived cells is an area of active investigation. Suratt and colleagues [67] were able to demonstrate up to 8% alveolar epithelial cell chimerism in three female subjects who had received gender mismatched stem cells, and a much higher percentage of pulmonary endothelial cells (up to 42%). In contrast to this, however, Kleeberger and colleagues were not able to demonstrate epithelial chimerism in bronchial tissue samples of three subjects [68], and Zander and colleagues could only demonstrate 1 male AEII cell in 3 PBSC transplant recipients [69]. Importantly, not all of these transplant subjects had evidence of on-going ALI. Given the need for lung injury to produce measurable engraftment in animals, these findings are not surprising. It likely is more difficult to identify donor cells in situations of lung homeostasis than it would be in response to more pronounced injury, such as in ALI.

A few human studies to date have demonstrated that the mere enumeration of circulating progenitor cells may have prognostic importance in ALI [70], and their numbers in circulation are directly related to proper lung repair in pneumonia, a common risk factor for ALI [71]. These studies are encouraging in that the bone marrow appears to be responding to lung injury, and suggest that further clinical studies may be worthwhile to further delineate what types of cells are integral to repair of the damaged lung.

CONCLUSION

The role of stem cells in repair of lung disorders such as ALI and ARDS is an active topic for current investigation, as indicated by recent NIH initiatives [14]. Animal studies suggest potential roles for both intrinsic lung stem cells as well as bone marrow-derived stem cells in the repair of the lung. Evidence exists that alveolar epithelial repair and pulmonary vascular endothelial repair may be related to the actions of these immature cells, or from more differentiated cells modifying their phenotype and thereby their activity. The development of high quality animal models with cutting-edge techniques of analysis will allow translational investigations in this area to proceed, with information gained leading to clinical trials.

Although stem cell therapy for ALI is certainly in the distant future, it is worthwhile pursuing further investigations given the lack of currently available therapies to help patients with this disorder.

ACKNOWLEDGMENTS

The authors wish to thank Meredith Mealer for her helpful reviews of the manuscript.

REFERENCES

[1] Ware LB, Matthay MA. The acute respiratory distress syndrome. *N Engl J Med* 2000; 342(18):1334-1349.

[2] Gattinoni L, Caironi P, Cressoni M et al. Lung recruitment in patients with the acute respiratory distress syndrome. *N Engl J Med* 2006; 354(17):1775-1786.

[3] Steinberg KP, Hudson LD, Goodman RB et al. Efficacy and safety of corticosteroids for persistent acute respiratory distress syndrome. *N Engl J Med* 2006; 354(16):1671-1684.

[4] Spragg RG, Lewis JF, Walmrath HD et al. Effect of recombinant surfactant protein C-based surfactant on the acute respiratory distress syndrome. *N Engl J Med* 2004; 351(9):884-892.

[5] Ventilation with lower tidal volumes as compared with traditional tidal volumes for acute lung injury and the acute respiratory distress syndrome. The Acute Respiratory Distress Syndrome Network. *N Engl J Med* 2000; 342(18):1301-1308.

[6] Uhal BD. Cell cycle kinetics in the alveolar epithelium. *Am J Physiol* 1997; 272(6 Pt 1):L1031-L1045.

[7] Griffiths MJ, Bonnet D, Janes SM. Stem cells of the alveolar epithelium. *Lancet* 2005; 366(9481):249-260.

[8] Ware LB, Eisner MD, Thompson BT, Parsons PE, Matthay MA. Significance of von Willebrand factor in septic and nonseptic patients with acute lung injury. *Am J Respir Crit Care Med* 2004; 170(7):766-772.

[9] Hristov M, Erl W, Weber PC. Endothelial progenitor cells: mobilization, differentiation, and homing. *Arterioscler Thromb Vasc Biol* 2003; 23(7):1185-1189.

[10] Asahara T, Murohara T, Sullivan A et al. Isolation of putative progenitor endothelial cells for angiogenesis. *Science* 1997; 275(5302):964-967.

[11] Moloney ED, Evans TW. Pathophysiology and pharmacological treatment of pulmonary hypertension in acute respiratory distress syndrome. *Eur Respir J* 2003; 21(4):720-727.

[12] Taylor RW, Zimmerman JL, Dellinger RP et al. Low-dose inhaled nitric oxide in patients with acute lung injury: a randomized controlled trial. *JAMA* 2004; 291(13):1603-1609.

[13] Matthay MA, Zimmerman GA, Esmon C et al. Future research directions in acute lung injury: summary of a National Heart, Lung, and Blood Institute working group. *Am J Respir Crit Care Med* 2003; 167(7):1027-1035.

[14] Weiss DJ, Berberich MA, Borok Z et al. Adult stem cells, lung biology, and lung disease. *NHLBI/Cystic Fibrosis Foundation Workshop. Proc Am Thorac Soc* 2006; 3(3):193-207.

[15] Welt FG, Losordo DW. Cell therapy for acute myocardial infarction: curb your enthusiasm? *Circulation* 2006; 113(10):1272-1274.

[16] Assmus B, Honold J, Schachinger V et al. Transcoronary transplantation of progenitor cells after myocardial infarction. *N Engl J Med* 2006; 355(12):1222-1232.

[17] Schachinger V, Erbs S, Elsasser A et al. Intracoronary bone marrow-derived progenitor cells in acute myocardial infarction. *N Engl J Med* 2006; 355(12):1210-1221.

[18] Lunde K, Solheim S, Aakhus S et al. Intracoronary injection of mononuclear bone marrow cells in acute myocardial infarction.*N Engl J Med* 2006; 355(12):1199-1209.

[19] Bernard GR, Artigas A, Brigham KL et al. The American-European Consensus Conference on ARDS. Definitions, mechanisms, relevant outcomes, and clinical trial coordination. *Am J Respir Crit Care Med* 1994; 149(3 Pt 1):818-824.

[20] Borok Z, Li C, Liebler J, Aghamohammadi N, Londhe VA, Minoo P. Developmental pathways and specification of intrapulmonary stem cells. *Pediatr Res* 2006; 59(4 Pt 2):84R-93R.

[21] Yen CC, Yang SH, Lin CY, Chen CM. Stem cells in the lung parenchyma and prospects for lung injury therapy. *Eur J Clin Invest* 2006; 36(5):310-319.

[22] Aliotta JM, Passero M, Meharg J et al. Stem cells and pulmonary metamorphosis: new concepts in repair and regeneration. *J Cell Physiol* 2005; 204(3):725-741.

[23] Kim CF, Jackson EL, Woolfenden AE et al. Identification of bronchioalveolar stem cells in normal lung and lung cancer. *Cell* 2005; 121(6):823-835.

[24] Kapanci Y, Weibel ER, Kaplan HP, Robinson FR. Pathogenesis and reversibility of the pulmonary lesions of oxygen toxicity in monkeys. II. Ultrastructural and morphometric studies. *Lab Invest* 1969; 20(1):101-118.

[25] Evans MJ, Cabral LJ, Stephens RJ, Freeman G. Transformation of alveolar type 2 cells to type 1 cells following exposure to NO2. *Exp Mol Pathol* 1975; 22(1):142-150.

[26] Shannon JM, Jennings SD, Nielsen LD. Modulation of alveolar type II cell differentiated function in vitro. *Am J Physiol* 1992; 262(4 Pt 1):L427-L436.

[27] Summer R, Kotton DN, Sun X, Ma B, Fitzsimmons K, Fine A. Side population cells and Bcrp1 expression in lung. *Am J Physiol Lung Cell Mol Physiol* 2003; 285(1):L97-104.

[28] Summer R, Kotton DN, Sun X, Fitzsimmons K, Fine A. Translational physiology: origin and phenotype of lung side population cells. *Am J Physiol Lung Cell Mol Physiol* 2004; 287(3):L477-L483.

[29] Majka SM, Beutz MA, Hagen M, Izzo AA, Voelkel N, Helm KM. Identification of novel resident pulmonary stem cells: form and function of the lung side population. *Stem Cells* 2005; 23(8):1073-1081.

[30] Krause DS, Theise ND, Collector MI et al. Multi-organ, multi-lineage engraftment by a single bone marrow-derived stem cell. *Cell* 2001; 105(3):369-377.

[31] Theise ND, Henegariu O, Grove J et al. Radiation pneumonitis in mice: a severe injury model for pneumocyte engraftment from bone marrow. *Exp Hematol* 2002; 30(11):1333-1338.

[32] Ortiz LA, Gambelli F, McBride C et al. Mesenchymal stem cell engraftment in lung is enhanced in response to bleomycin exposure and ameliorates its fibrotic effects. *Proc Natl Acad Sci U S A* 2003; 100(14):8407-8411.

[33] Abe S, Boyer C, Liu X et al. Cells derived from the circulation contribute to the repair of lung injury. *Am J Respir Crit Care Med* 2004; 170(11):1158-1163.

[34] Yamada M, Kubo H, Kobayashi S et al. Bone marrow-derived progenitor cells are important for lung repair after lipopolysaccharide-induced lung injury. *J Immunol* 2004; 172(2):1266-1272.

[35] Prockop DJ. Marrow stromal cells as stem cells for nonhematopoietic tissues. *Science* 1997; 276(5309):71-74.

[36] Friedenstein AJ, Gorskaja JF, Kulagina NN. Fibroblast precursors in normal and irradiated mouse hematopoietic organs. *Exp Hematol* 1976; 4(5):267-274.

[37] Pittenger MF, Mackay AM, Beck SC et al. Multilineage potential of adult human mesenchymal stem cells. *Science* 1999; 284(5411):143-147.

[38] Rojas M, Xu J, Woods CR et al. Bone marrow-derived mesenchymal stem cells in repair of the injured lung. *Am J Respir Cell Mol Biol* 2005; 33(2):145-152.

[39] Jiang Y, Jahagirdar BN, Reinhardt RL et al. Pluripotency of mesenchymal stem cells derived from adult marrow. *Nature* 2002; 418(6893):41-49.

[40] Reyes M, Lund T, Lenvik T, Aguiar D, Koodie L, Verfaillie CM. Purification and ex vivo expansion of postnatal human marrow mesodermal progenitor cells. *Blood* 2001; 98(9):2615-2625.

[41] Reyes M, Dudek A, Jahagirdar B, Koodie L, Marker PH, Verfaillie CM. Origin of endothelial progenitors in human postnatal bone marrow. *J Clin Invest* 2002; 109(3):337-346.

[42] Hristov M, Erl W, Weber PC. Endothelial progenitor cells: isolation and characterization. *Trends Cardiovasc Med* 2003; 13(5):201-206.

[43] Hill JM, Zalos G, Halcox JP et al. Circulating endothelial progenitor cells, vascular function, and cardiovascular risk. *N Engl J Med* 2003; 348(7):593-600.

[44] Reddy R, Buckley S, Doerken M et al. Isolation of a putative progenitor subpopulation of alveolar epithelial type 2 cells. *Am J Physiol Lung Cell Mol Physiol* 2004; 286(4):L658-L667.

[45] Kotton DN, Ma BY, Cardoso WV et al. Bone marrow-derived cells as progenitors of lung alveolar epithelium. *Development* 2001; 128(24):5181-5188.

[46] Summer R, Kotton DN, Liang S, Fitzsimmons K, Sun X, Fine A. Embryonic lung side population cells are hematopoietic and vascular precursors. *Am J Respir Cell Mol Biol* 2005; 33(1):32-40.

[47] Francois S, Bensidhoum M, Mouiseddine M et al. Local irradiation not only induces homing of human mesenchymal stem cells at exposed sites but promotes their widespread engraftment to multiple organs: a study of their quantitative distribution after irradiation damage. *Stem Cells* 2006; 24(4):1020-1029.

[48] Lapidot T, Petit I. Current understanding of stem cell mobilization: the roles of chemokines, proteolytic enzymes, adhesion molecules, cytokines, and stromal cells. *Exp Hematol* 2002; 30(9):973-981.

[49] Fuste B, Escolar G, Marin P, Mazzara R, Ordinas A, az-Ricart M. G-CSF increases the expression of VCAM-1 on stromal cells promoting the adhesion of CD34+ hematopoietic cells: studies under flow conditions. *Exp Hematol* 2004; 32(8):765-772.

[50] Fruehauf S, Seggewiss R. It's moving day: factors affecting peripheral blood stem cell mobilization and strategies for improvement [corrected]. *Br J Haematol* 2003; 122(3):360-375.

[51] Aicher A, Zeiher AM, Dimmeler S. Mobilizing endothelial progenitor cells. *Hypertension* 2005; 45(3):321-325.

[52] Aiuti A, Webb IJ, Bleul C, Springer T, Gutierrez-Ramos JC. The chemokine SDF-1 is a chemoattractant for human CD34+ hematopoietic progenitor cells and provides a new mechanism to explain the mobilization of CD34+ progenitors to peripheral blood. *J Exp Med* 1997; 185(1):111-120.

[53] Heeschen C, Aicher A, Lehmann R et al. Erythropoietin is a potent physiologic stimulus for endothelial progenitor cell mobilization. *Blood* 2003; 102(4):1340-1346.

[54] Bahlmann FH, de GK, Spandau JM et al. Erythropoietin regulates endothelial progenitor cells. *Blood* 2004; 103(3):921-926.

[55] Brigham KL, Stecenko AA. Gene therapy for acute lung injury. *Intensive Care Med* 2000; 26 Suppl 1:S119-S123.

[56] Nagaya N, Kangawa K, Kanda M et al. Hybrid cell-gene therapy for pulmonary hypertension based on phagocytosing action of endothelial progenitor cells. *Circulation* 2003; 108(7):889-895.

[57] Zhao YD, Courtman DW, Deng Y, Kugathasan L, Zhang Q, Stewart DJ. Rescue of monocrotaline-induced pulmonary arterial hypertension using bone marrow-derived endothelial-like progenitor cells: efficacy of combined cell and eNOS gene therapy in established disease. *Circ Res* 2005; 96(4):442-450.

[58] Kanki-Horimoto S, Horimoto H, Mieno S et al. Implantation of mesenchymal stem cells overexpressing endothelial nitric oxide synthase improves right ventricular impairments caused by pulmonary hypertension. *Circulation* 2006; 114(1 Suppl):I181-I185.

[59] Aliotta JM, Keaney P, Passero M et al. Bone marrow production of lung cells: the impact of G-CSF, cardiotoxin, graded doses of irradiation, and subpopulation phenotype. *Exp Hematol* 2006; 34(2):230-241.

[60] Gross NJ. Surfactant subtypes in experimental lung damage: radiation pneumonitis. *Am J Physiol* 1991; 260(4 Pt 1):L302-L310.

[61] Wang SL, Liao Z, Vaporciyan AA et al. Investigation of clinical and dosimetric factors associated with postoperative pulmonary complications in esophageal cancer patients treated with concurrent chemoradiotherapy followed by surgery. *Int J Radiat Oncol Biol Phys* 2006; 64(3):692-699.

[62] Tashiro K, Yamada K, Li WZ, Matsumoto Y, Kobayashi T. Aerosolized and instilled surfactant therapies for acute lung injury caused by intratracheal endotoxin in rats. *Crit Care Med* 1996; 24(3):488-494.

[63] West JA, Pakehham G, Morin D, Fleschner CA, Buckpitt AR, Plopper CG. Inhaled naphthalene causes dose dependent Clara cell cytotoxicity in mice but not in rats. *Toxicol Appl Pharmacol* 2001; 173(2):114-119.

[64] Beckett T, Loi R, Prenovitz R et al. Acute lung injury with endotoxin or NO2 does not enhance development of airway epithelium from bone marrow. *Mol Ther* 2005; 12(4):680-686.

[65] Chien KR. Stem cells: lost in translation. *Nature* 2004; 428(6983):607-608.

[66] Kotton DN, Fabian AJ, Mulligan RC. Failure of bone marrow to reconstitute lung epithelium. *Am J Respir Cell Mol Biol* 2005; 33(4):328-334.

[67] Suratt BT, Cool CD, Serls AE et al. Human pulmonary chimerism after hematopoietic stem cell transplantation. *Am J Respir Crit Care Med* 2003; 168(3):318-322.

[68] Kleeberger W, Versmold A, Rothamel T et al. Increased chimerism of bronchial and alveolar epithelium in human lung allografts undergoing chronic injury. *Am J Pathol* 2003; 162(5):1487-1494.

[69] Zander DS, Cogle CR, Theise ND, Crawford JM. Donor-derived type II pneumocytes are rare in the lungs of allogeneic hematopoietic cell transplant recipients. *Ann Clin Lab Sci* 2006; 36(1):47-52.

[70] Burnham EL, Taylor WR, Quyyumi AA, Rojas M, Brigham KL, Moss M. Increased circulating endothelial progenitor cells are associated with survival in acute lung injury. *Am J Respir Crit Care Med* 2005; 172(7):854-860.

[71] Yamada M, Kubo H, Ishizawa K, Kobayashi S, Shinkawa M, Sasaki H. Increased circulating endothelial progenitor cells in patients with bacterial pneumonia: evidence that bone marrow derived cells contribute to lung repair. *Thorax* 2005; 60(5):410-413.

INDEX

A

abortion, 60, 64, 69, 73
accuracy, 72
acetylcholine, 148
acid, 4, 5, 7, 35, 98, 142, 154
acidic, viii, 31, 45
activation, 16, 85, 93, 118, 119, 120, 125, 126, 128, 133, 134, 152, 160
acute lung injury, x, 159, 160, 175, 178, 179
acute myeloid leukemia, 15
acute respiratory distress syndrome, x, 159, 160, 175
adenovirus, 93, 171
adhesion, 29, 99, 113, 142, 165, 168, 169, 177
adipocyte(s), viii, 31, 32, 36, 44, 47, 96, 97, 98, 102, 103, 104, 105, 107, 108, 114, 115, 116
adipogenic, viii, 31, 32, 34, 36, 37, 38, 44, 97, 98, 99, 101, 104, 107, 108, 114
adipose, viii, x, 31, 32, 33, 36, 37, 44, 45, 46, 47, 48, 49, 50, 95, 96, 97, 99, 100, 102, 103, 104, 105, 107, 108, 109, 110, 111, 112, 113, 115, 116, 164
adipose tissue, viii, x, 31, 32, 33, 36, 44, 46, 50, 95, 96, 97, 99, 100, 102, 103, 104, 105, 107, 108, 109, 110, 111, 112, 113, 115, 116, 164
administration, 127, 131, 167, 169, 171, 172
adult organisms, vii
adult stem cells, vii, viii, 47, 48, 51, 54, 56, 113
adult tissues, vii, 52, 54, 55, 96, 104
adulthood, 52
adults, 67, 69, 70, 71, 75, 102, 104, 121
age(ing), 33, 58, 68, 69, 112, 121, 149
agent, 73, 167, 169, 172, 173
aggregates, 35, 45, 145
airways, 163
albinism, 144
algorithm, 151
ALI, x, 159, 160, 161, 162, 163, 166, 168, 169, 170, 171, 172, 173, 174
alkaline, viii, 31, 35, 40

alkaline phosphatase (ALP), viii, 31, 35, 37, 45, 40
alternative(s), vii, x, 1, 65, 95, 96, 108, 115, 133, 157, 169, 172
alveolus(i), 160, 162, 163, 166, 173
Alzheimer's disease, vii
amino acids, 83, 84
AMR, 157
anatomy, 20
anemia, 29
angiogenesis, 96, 101, 110, 114, 175
animal models, xi, 142, 159, 161, 169, 170, 172, 173, 174
animals, 13, 61, 73, 76, 90, 148, 165, 166, 167, 168, 169, 170, 172, 173, 174
antibiotics, 161
antibody(ies), 21, 27, 83, 86, 87, 113, 165, 173
antigen, x, 16, 21, 22, 27, 117, 118, 120, 121, 124, 126, 127, 128, 130, 132, 134, 163, 164
anus, 138, 139
aorta, 103, 105, 115
APC, 126
aplasia, 2
apoptosis, 16, 133
apoptotic cells, 126
appendix, 78
ARDS, x, 159, 160, 161, 162, 172, 173, 174, 176
argument, 46, 61, 62, 63, 66, 68, 69, 70, 71, 73
Aristotle, 63, 71, 72, 76, 78
arrest, 13
arterial hypertension, 178
artery(ies), 99, 102, 103, 114, 170
articular cartilage, 50
aspiration, viii, 19, 20, 21, 22, 23, 24, 27, 29, 169
assessment, 10, 13, 14
astrocytes, 53, 54, 55, 56
attachment, 160
attention, ix, x, 59, 60, 142, 159, 161, 172
Australia, 117
authority, 63
autoimmune diseases, vii

D

E

N

O

P